OXFORD HISTORICAL MONOGRAPHS

Editors

REALISTIC UTOPIAS
The Ideal Imaginary Societies
of the Renaissance
1516-1630

BY

MIRIAM ELIAV-FELDON

CLARENDON PRESS · OXFORD
1982

Oxford University Press, Walton Street, Oxford OX2 6DP

LONDON   GLASGOW   NEW YORK   TORONTO
DELHI   BOMBAY   CALCUTTA   MADRAS   KARACHI
KUALA LUMPUR   SINGAPORE   HONG KONG   TOKYO
NAIROBI   DAR ES SALAAM   CAPE TOWN
MELBOURNE   AUCKLAND
and associates
BEIRUT   BERLIN   IBADAN   MEXICO CITY   NICOSIA

Published in the United States by
Oxford University Press, New York

British Library Cataloguing in Publication Data

Eliav-Feldon, Miriam
    Realistic utopias.—(Oxford historical monographs)
    1. Utopias—History
    I. Title
    321'.07   HX806

ISBN 0-19-821889-3

Library of Congress Cataloging in Publication Data

Eliav-Feldon, Miriam, 1946–
    Realistic utopias.
    (Oxford historical monographs)
    Bibliography: p.
    Includes index.
    1. Utopias—History. I. Title.
HX806.E45 1982   321'.07   82-7975
ISBN 0-19-821889-3 (Oxford University Press)
                                                   AACR2

Typeset and printed by The Thetford Press Ltd., Thetford, Norfolk

To the memory of my parents
Berta and Binyamin Eliav

ACKNOWLEDGEMENTS

I should like to express my deepest gratitude to Professor H. R. Trevor-Roper (now Lord Dacre of Glanton) whose advice, guidance, and encouragement sustained me throughout three years of research and writing.

For financial support I am greatly indebted to the Anglo-Israel Association for awarding me the Kenneth Lindsay Scholarship which enabled me to pursue my studies in Oxford in the years 1974-7; and to the Hebrew University of Jerusalem for supplementary annual awards.

I also wish to thank Sir Isaiah Berlin who was most helpful in my early stages at Oxford. To Wolfson College — its President, members, and staff — I am grateful for the pleasant atmosphere and facilities which made my stay in Oxford such an enjoyable experience. And to Joram Feldon—for everything.

CONTENTS

1

INTRODUCTION

A utopia is an invitation to perceive the distance between things as they are and things as they should be. It is a presentation of a positive and possible alternative to the social reality, intended as a model to be emulated or aspired to. Since it is an appeal to perfect the social environment, it expresses explicit and implicit criticism of the things as they are. The first phenomena to be abolished in utopia are obviously those that the author considers to be the major evils of his own society. 'I'the commonwealth I would *by contraries* Execute all things.'[1] Unlike other proposals for reform, a utopia depicts an entire and functioning society, and thus it becomes a prism through which is visible the entire spectrum of the author's feelings about the society that surrounds him with its institutions, laws, customs, and idiosyncrasies. Together with the anguish and the censure, a utopia contains the hopes and the beliefs in the possibilities open to contemporary society. A group of utopias from the same period forms, therefore, an excellent collection of documents that offer a mosaic of social thoughts and feelings. Or, in the words of Lucien Febvre: 'Ils parlent quand l'humanité, inquiète, cherche à préciser les grandes lignes de boulversements sociaux et moraux, que chacun sent inévitables et menaçants. Par-là leur oeuvres sont, pour l'historien, des témoignages souvent pathétiques, toujours intéressants, non pas seulment de la fantaisie et de la imagination de quelques précurseurs, mais de l'état intime d'une societé.'[2]

Utopias as a rule have little literary value, despite their fictional framework. Nor are they very valuable as sources of information about the *facts* of the historical reality, especially when the period under discussion is rich with other documentary material. It is very doubtful, for instance, whether one could learn anything new about the extent of enclosures in England in the sixteenth century from the utopias of Thomas More and Robert Burton. However, they are certainly

---

[1] Gonzalo in *The Tempest*, II. i. 154-5 (italics added).
[2] Lucien Febvre, *Pour une histoire à part entière*, Paris, 1962, pp. 741-2.

important as sources of information about the opinions and feelings of contemporaries concerning these facts and the problems of their age.

Genuine utopists do not indulge in fantasies about unattainable Gardens of Eden, but propose practical, though sometimes very drastic remedies for the defects of their societies. The method of describing a functioning society, with all its components and details, only emphasizes their practical and realistic intentions: the fabricated authenticity makes ideas into concrete, believable 'facts'. The unique trait of utopias as documents of social theory is that in them the concrete explains the abstract, the vivid picture lends attraction to the theory. A secondary point, but a very important one at certain periods, is that the method of disguising one's political and social censure in the form of an imaginary land had the added advantage of protecting the author from the wrath of the censor.

Utopias are written by intellectuals who are sensitive to the misery of this world. A utopia is an expression of malaise, though rarely that arising from personal suffering, but rather from aesthetic discontent in view of squalor, corruption, disorder, and disharmony. This aesthetic malaise is by no means confined to any one period — intellectuals of all times have suffered from it. A historian reading utopias as social documents must learn, therefore, to disentangle the specific from the universal, as well as to remember that he is offered in fact a view of society from aside or from above. However, these disadvantages are compensated for by the power of expression, the totality of the vision, and the penetrating criticism that many of the utopias were capable of offering.

A vast number of definitions had been given to the word 'utopia' since it was coined by Thomas More. To repeat them all here would serve no purpose. In this study 'utopia' will be used in a rather narrow sense to mean only *a literary work describing an ideal society created by conscious human effort on this earth*. This definition excludes any vision of an ideal existence that is other-worldly, unattainable, or dependent either on wonders of nature or on divine intervention. It also excludes states of bliss of one individual or other forms of personal salvation, as well as recommendations for piecemeal reforms and schemes for the amelioration of the human condition in any one particular field. But even this exclusive definition still applies to a wide variety of works which could be tentatively grouped as follows:[3]

[3] This classification is based only on the types of utopian works known in the Renaissance. All the examples are taken from that period. It is not, perhaps, an exhaustive division, and some groups may overlap. It is mainly intended for the purpose of seeing the group of utopias that was selected for this study in the context of other works of the same or of a similar genre.

(1) *The good government:* Constitutions for and speculations upon the best commonwealth, mirrors for princes, practical political recommendations. Such works usually lack a fictional framework and the vivid picture of a living society. They are more in the nature of advice to the Good Prince — a form that proliferated during the Renaissance. To this group belong works such as Antonio de Guevara, *Libro Llamado Relox de los Principes* (1529); Alfonso de Valdés, *Dialogo de Mercurio y Carón* (1529); Thomas Starkey, *Dialogue between Cardinal Pole and Thomas Lupset* (1536-8); Martin Bucer, *De Regno Christi* (1557); and Giovanni Bonifacio, *La repubblica delle api* (1627).[4]

(2) *Idealizations of existing societies:* These are to be found in travellers' tales or in expositions of certain long-standing myths such as the admiration for the polity of Venice. They are also meant to serve as exemplary models, but since they are not entirely products of the author's imagination, they are not presented as faultless. Among them one could number Cardinal Gasparo Contarini, *De magistratibus et republica Venetorum* (1543); Juan de Salazar, *Politica Española* (1619); and Ludovico Zuccolo's admiration for the republic of San Marino in *La Città Felice* (1625).

(3) *Designs of ideal cities:* A form popular in Quattrocento Italy. Many of these contain some social ideas, but the main emphasis is on architectural planning. Or, in other words, these are designs for the 'container' — the physical environment — rather than a programme for the 'content', i.e. the human beings that comprise a society. Among them are Leon Battista Alberti, *De re aedificatoria* (1452); Filarete (Antonio Averlino) *Sforzinda* (1464); and Leonardo da Vinci's plan (1487) that was first published in *Manuscrit B de L'Institut de France* (1960).

(4) *Glorifications of a primitive Golden Age:* A paradise lost of innocence and of life according to nature, sometimes in a mythological past, sometimes in the New World before the Europeans came and destroyed it. But it is an unredeemable paradise, and therefore it is a lamentation rather than a model to be emulated. Some examples would be the anonymous *La Pazzia* (*c*.1541): Giordano Bruno, *Spaccio della bestia trionfante* (1584); Marco Girolamo Vida, *Dialogi de rei publicae dignitate* (1556); and Garcilaso de la Vega, *Primera parte de los commentarios reales* (1609).

---

[4] Bonifacio's work belongs, in fact, to two different categories: it begins with a description of a primitive society of happy people without letters or laws (group 4), and then it offers a constitution to be imposed on these people to make them into a civilized Christian nation (group 1). The constitution is based on Virgil's laws of the bees. On the frequent use of beehive and anthill imagery in utopian writing see M. J. Lasky, *Utopia and Revolution,* Chicago and London, 1976, pp. 4-8.

(5) *Secret societies (real or imaginary):* A select élite that aspires to reform Christendom, or even the entire world. Programmes for or constitutions of such organizations are based on the assumption that all members are virtuous and thus dispense with the need to solve the mundane problems of control and regulation. Such are the works of Francesco Pucci, *Forma d'una repubblica cattolica* (1581); Johann Valentin Andreae, *Christianae Societatis Imago* (1619); Johannes Amos Comenius, *The Labyrinth of the World and the Paradise of the Heart* (1623).

(6) *World empires and plans for universal eternal peace:* These are perhaps the only true utopias, since only in them is abolished the evil of war which menaces all other utopian lands. However, these works devote little attention to internal social problems. Among them are the works of Guillaume Postel, *Concordia Mundi* (1548); Tommaso Campanella, *Monarchia di Spagna* (1599); Ottavio Pallavicino, *La Repubblica Cristiana* (*c.*1605-21); Émeric Crucé, *Le Nouveau Cynée* (1623).

(7) *Theocratic millennial kingdoms (plans or experiments):* Attempts to found ideal societies in the New World were made by members of the monastic orders. Two survived for a considerable time: Vasco de Quiroga's Pueblo Hospitals in Santa Fé, and the Indian settlements established by the Jesuits in Paraguay. Another such plan, by the Franciscan Geronimo de Mendieta, was never realized. Although directly influenced by the utopias of the humanists, these experiments were more in line of the myth of Antilia — a blessed theocracy governed by the bishops.

In Europe itself the obvious example is the revolutionary attempt of the Anabaptists in Münster; and in the literary field it is *Das Buch der Hundret Kapitel und der Vierzig Statuten des sogennanten Oberrheinischen Revolutionärs* (The Book of a Hundred Chapters by the Revolutionary of the Upper Rhine (1510)).[5]

(8) *Utopias proper:* Ideal imaginary societies described in their entirety as if functioning in the present. Many of them make use of a fictional narrative about a voyage across the ocean and the discovery of the happy society that had achieved social perfection in the past and retained it ever since, on some distant shores — usually on an island. But some authors

---

[5] Many works mentioned here are listed, and briefly described, in the bibliography of Utopiana compiled by R. W. Gibson and J. Max Patrick in R. W. Gibson *St. Thomas More: A Preliminary Bibliography,* Yale University Press, 1961. Two useful recent bibliographies are Glenn Negley, *Utopian Literature: A Bibliography with a Supplementary Listing of Works Influential in Utopian Thought,* Lawrence: Regents Press of Kansas, 1978; and L. T. Sargent, *British and American Utopian Literature 1516-1975,* Boston, 1979. A few of the works mentioned in groups (1)-(7) were not easily available, and my information on them is based on secondary sources (see Bibliography).

did not strain their imagination to distant lands, and instead of recounting a story about a voyage, they present their perfect societies as philosophical speculation in dialogue form.

The present study is devoted to this group only, and hereafter the term 'utopia' is applied only to this category. It consists of the following works:

1. Thomas More, *Utopia* (1516)
2. Johan Eberlin von Günzburg, *Wolfaria* (1521)
3. François Rabelais, 'L'Abbaye des Thélémites' (1534)
4. Francesco Patrizi da Cherso, *La Città Felice* (1551)
5. Anton Francesco Doni, 'Il Mondo Savio e Pazzo' (1552)
6. Gasparus Stiblinus, *De Eudaemonensium Republica* (1553)
7. Ludovico Agostini, *Repubblica Immaginaria* (1585-90)
8. Tommaso Campanella, *La Città del Sole* (1602)
9. I. D. M., *Le Royaume d'Antangil* (1616)
10. Johann Valentin Andreae, *Christianopolis* (1619)
11. Robert Burton, 'A Utopia of Mine Own' (1621)
12. Francis Bacon, *New Atlantis* (1624)
13. Ludovico Zuccolo, *La Repubblica d'Evandria* (1625).[6]

In 1516 the genre of genuine secular utopias was reborn after a complete absence since Hellenic times. Some very important obstacles had to be transcended before this type of social thought could reappear. First, utopias could not be written as long as the existing structure of society was considered to be divinely ordained. One was not supposed to reflect on the wishes of God and on His creation: if this is how things are, then this is how they ought to be. Of course, one admitted that there were many social evils — but they were punishments visited upon sinful society. And the only hope was repentance, a change of heart, that would bring divine forgiveness and the abolition of evil, which may even come about in some dramatic, apocalyptic manner. Utopias proper were possible only when some men began to feel that human beings were responsible for the existing deficiencies, and therefore it was within their power to correct them.

Second, utopias became possible only when the *contemptus mundi* was rejected. When life on this earth was no longer regarded as a dark vestibule leading to the great hall of the next world, only then was the right of man to happiness on this earth recognized.

Third, utopias were not possible as long as the dominant belief was in the constant degeneration of humanity since the Fall. While this belief

---

[6] More details about each author and each work are given in section B of this introduction.

prevailed, ideas of an ideal life were projected either into a mythological past, a Golden Age, or into the next world, to the afterlife in dreams of Heaven or of a *Civitas Dei*.

These basic changes in the analysis of the human condition could occur only when there came into existence a generation of lay intellectuals who were not fettered by the concepts of theology.[7]

The very same obstacles stood in the way of the idea of progress. But the fact that both these philosophical phenomena required the same changes in thought does not necessarily mean that the roots of the idea of progress are concealed in the utopias. On the contrary: a utopia is a static society that has not evolved gradually but was created by fiat *ex nihilo*, has not changed since its creation, and is not destined to change in the future, since any change means deviation from perfection and therefore corruption. It is also a society artificially cut off from the course of human history.

The utopias of the Renaissance in no way express a belief in the future, but at the same time they are no longer a lamentation of the past. All the utopias that are the subject of this study are described as existing or as possible in the present. The authority of the ancients had been undermined by the great inventions and discoveries of the age, and these also gave contemporaries the confidence that they were not inferior in any way to classical times. They were conscious of the new, quickening pace of history, and extremely proud of it: '. . . our age has more history in it in a hundred years than the whole world in the preceding 4,000 years. . . .'.[8] But they did not make the further step of assuming that if the present is better than the past, then the future will be better than the present.

Unlike later creations of the same genre, the Renaissance utopias do not express the belief that science and technology are means to a better civilization. Even in Bacon's New Atlantis, where the College of Science is 'the lantern of the kingdom', science has made no apparent contribution to the improvement of society. On the other hand, if science is not yet regarded as an aid towards progress, it is also no longer considered to be the sphere of Satan. The study of nature is emphatically encouraged.

---

[7] Another test-case which, I believe, supports the argument that utopias proper are possible only within some form of a *secular* culture is that of the Zionist utopias in the second half of the nineteenth century. Before that time there were no Jewish utopias but only expectations for the coming of the Messiah, The first genuine Jewish utopias appear simultaneously with the first generation of non-rabbinical, yet non-assimilationists, Jewish men of letters.

[8] Campanella, 'The City of the Sun', in G. Negley and J. Max Patrick, *The Quest for Utopia*, New York, 1952, p. 345.

As with the idea of progress, the roots of primitivism and the romanticization of the Noble Savage grew parallel to the evolution of utopian thought, also as a consequence of the same changes in the prevailing philosophy and the impact of the New World. But again, the utopias themselves are not primitivist at all. Undoubtedly, some of the authors were impressed by information about the inhabitants of the New World, but their ideal societies are highly sophisticated, urban, as 'modern' as they could possibly be, and not lacking any of the achievements of European civilization.

The utopias of the Renaissance are in many respects a watershed: a curious mixture of optimism and pessimism, the sense of a new vigour, a new vitality, and yet of helplessness and impotence to effect real changes. They were already an expression of the awareness that, in order to create a good society, drastic changes in social structure and in institutions were required, and yet they are not plans for action. They do not describe the bridge that would lead from the condemned reality to the ideal model. Although the utopists were aware that it is not sufficient to rely solely on the quality of the rulers, they could only pin their hopes on enlightened princes who would heed the advice of philosophers, because, despite the rehabilitation of Man in humanist thought, most of these lay intellectuals felt only contempt and suspicion for the multitude, for real men. Simple men and women were to be the building-blocks of the utopia, but not, as a group or a movement, its builders.

Many scholars[9] claim that every period of a major transition is marked by an efflorescence of utopian writing, since utopias are the best vehicle to express both the deep fears and the high hopes which are felt at each of the turning-points in history. This seems to be a position that is slightly difficult to maintain. In fact, the stream of utopias has continued constantly from the time of Thomas More until the horrors and the great disillusions of the twentieth century had turned utopias sour. And if utopias are in evidence throughout at least four centuries, it seems rather meaningless to say that they mark periods of transition. It is probably only the sudden outburst of utopias in the sixteenth century that these scholars had in mind. Because, even regardless of their content, the mere fact of their appearance in this specific utopian form marks the gradual transition from the Middle Ages to the modern era, in social realities and in thought, knowledge, and attitudes towards man, society, church, government.

9 Such as Jean Servier, *Histoire de l'utopie*, Paris, 1967: Lucien Febvre, op cit.; Marie Louise Berneri, *Journey Through Utopia*, New York, 1971; and a few others.

'C'est l'homme qui joue à être dieu et non l'homme qui rêve d'un monde divin.'[10] A utopia is indeed the City of Man and not a City of God. And yet only from Rabelais's Abbey of Thélème is the hand of God banished completely. Faith, some form of ritual, and an ecclesiastical establishment exist in all other utopias — atheistic rationalism is still unthinkable. Despite all their 'modern' features, despite the clear break from medieval thought, it must be remembered that these utopias were written in a deeply religious environment and during the most dramatic religious upheavals in the history of Western Europe. Our thirteen utopias represent almost all shades of the religious spectrum of those tumultuous times: Christian humanism on the eve of the Reformation; the impact of the encounter with pagan civilizations; the zeal of revolutionaries immediately after the great storm broke out; Calvinist severity; state Anglicanism; the dogmatism and the strictness of Catholicism after Trent; the ironic irreverence born of the disillusion with both reformations; and last — heretical sectarianism.

The religious content of these works is not our concern in this study, and thus I shall not attempt any generalizations or answers to such difficult, and much discussed, questions as 'Why did Thomas More, the *hereticis molestus,* the Catholic martyr, describe in his ideal land a society of non-Christian polytheism and of toleration?' It is not directly relevant to this study whether More, together with Doni, Patrizi, and Campanella, described 'pagan' ideal societies because they were bound to do so in consequence of the fictional framework that made the utopia a land in the newly discovered pagan world, or because they wanted 'to shame' Europeans with the comparison. The point that needs to be emphasized is that, whether its religion is imaginary and non-Christian, or resembles one of the European churches minus its deficiencies, every utopia of the Renaissance is permeated with the religious aspirations, attitudes, criticism, and bias of the author. Together the utopists desired the reunification of Christendom, an end to religious strife, a true Reformation that would answer the spiritual needs of believers and would provide them with a true spiritual leadership. Some sought the rational, 'natural', foundations of Christianity as common ground for unification; some went so far as to advocate a rudimentary deism; others adhered to one of the existing churches. All criticized and condemned the corruption, the bigotry, and the negligence of the clergy. None was indifferent to religious issues.

[10] Raymond Ruyer, *L'Utopie et les utopies*, Paris, 1950, p. 4.

Real toleration was still an impossible notion, even for those who conceived lands of vague deism or polytheism. The main obstacle was the conviction that social morality would be impossible unless everybody retained at least the belief in the immortality of the soul and in reward and punishment in the afterlife. And it is the social role of religion that will be relevant to this discussion.

In their clamour for a total spiritual reformation the conflicts of their time are again apparent. On the one hand they professed a belief in the power of education and knowledge to mould better men and thus a better society; on the other hand they were obsessed by the fear that, like the builders of the Tower of Babel, humanity will suffer divine punishment for its *hubris,* and therefore they constantly sought measures to restrain it and to impose humility.

Although they are no longer satisfied with analysing social problems solely as sins and punishments, nevertheless they still use the terminology. The evils of the existing society and the advantages of the imaginary society are often contrasted as personifications of the Christian concepts of vices and virtues: Pride versus Humility, Avarice versus Frugality. What M. L. Berneri wrote about Andreae could be applied to most of his fellow utopists: ' . . . his love of men inclined him to trust them as sensible beings capable of going about their lives in a reliable and honest way, but his religion told him that man is wicked and has to be carefully guided, preached to and, if necessary, threatened, to be kept away from sin'.[11] This ambivalence afflicts, with a few exceptions, all our utopists in their attitude to each and every social issue: 'les utopies de la Renaissance admettent donc les prémises pessimistes du péché originel tout en refusant la salvation chrétienne'.[12]

Disaffection and criticism of the existing religious institutions were widespread phenomena in Renaissance Europe, but several of the utopists, together with a few of the composers of quasi-utopias in the late sixteenth and early seventeenth centuries, were also affected by the movement which Frances Yates has named 'the Rosicrucian Enlightenment': sects and individuals, branded as 'heretics' by all churches, tied by invisible links all over Europe, maintaining esoteric creeds, influenced by the 'Hermetic-cabalist, magico-scientific' tradition, were exhilarated by the sense of new powers and possibilities to change the human condition. Andreae was a major figure in that network; Campanella and Bacon, Pucci, Comenius, and Bruno were connected to it or influenced

---

[11] M. L. Berneri, op cit., p. 107.
[12] J. Servier, op. cit;. p. 155.

by its set of beliefs.[13] A great deal more detective work is required to reveal the entire scope of this phase in the history of European culture, and to uncover the place of utopias in it. For, when reading these works one is immediately struck by the many similarities in purpose, in hopes and excitements, between the describers of imaginary lands of social perfection and such elusive sects as the Familists and the Rosicrucians. But, with the horrors of the Thirty Years' War, with the suppression of the intellectual ferment in Italy, with the trial of Galileo, and in England the sounds of the approaching Civil War, comes a definite change in the tenor of the times. No definition of the Renaissance was ever stretched beyond the 1630s — the spirit of exhilaration, which was the product of the great transition, is gone. Thus, although the flow of utopian writing did not cease, the discussion of the utopias of the Renaissance will end with the works written in the third decade of the seventeenth century.

The utopia was, usually, only one small work among the many other writings by the same person or, sometimes, but a chapter in a voluminous work. The writer himself seemed, as a rule, to be slightly apologetic about his indulgence in fantasy; and we cannot determine today what was the attitude of his contemporaries to the utopia. (Thomas More is, of course, an exception in this as in many other respects.) Many of these utopias sank into oblivion quite soon after they were written; some of the utopists suffered the fate of their creation; others remained known for works and deeds unconnected with the utopias. For many generations no one paid much attention to these *jeux d'esprit*, except perhaps a few historians and students of the literature of the Renaissance who kept alive the memory of a small number of these works.

A pronounced change occurred in the second half of the nineteenth century. With the rise of Socialism as an ideology of mass movements, and the ferocious debates that it aroused, some of the long-forgotten works rose to the surface again. The Renaissance visions of communistic societies suddenly gain respectability and a wide public as 'the precursors of Socialism' — they are no longer regarded as mere

[13] See F. Yates, *The Rosicrucian Enlightenment*, London, 1972; id., *Giordano Bruno and the Hermetic Tradition*, London, 1964; L. Firpo, *Gli scritti di Francesco Pucci*, Turin, 1957; and D. Cantimori, *Eretici italiani del cinquecento*, Florence, 1967.

curiosities of interest only to the specializing historian.[14] True, this interest applied only to the 'revolutionary' utopias: More, Doni, Campanella, Andreae. The 'conservative' ideal societies remained practically unknown. Even when intellectual history became a popular discipline, and books began to be written on the history of utopian thought from the Bible to science fiction, the limelight was still only on the proposals for drastic rearrangements of society. The utopias even of quite well-known historical figures (Patrizi, Rabelais, Burton), not to mention the obscure or anonymous ones (Eberlin von Günzburg, Stiblinus, I.D.M.), who accepted the basic structure of society or wanted to restore even less egalitarian structures, did not attain a place in the anthologies until very recent times.

Nowadays many scholars realize that these descriptions of ideal societies are not simply pleasant day-dreams, but, in fact, the embodiment, the crystallization, of the entire *Weltanschauung* of the author. Thus their value as historical documents increased manifoldly. Now we are witnessing perhaps an opposite process: it is their writings on religion, philosophy, astrology, or other scientific subjects, and even the poetry and *belles lettres,* that are regarded as curiosities for historians only, while their visions of social perfection are regarded as interesting and relevant for a wide public.

In the best erudite and syncretic manner of the Renaissance, the utopists borrowed ideas from everywhere: from Plato, of course, and from the traditional myth of Sparta; from Aristotle and from the fragments of the Hellenic utopias; from the information that was reaching Europe about civilizations in other continents (the New World and China): and from existing examples — invariably romanticized — within the boundaries of Europe, such as Venice, Florence, Geneva, San Marino, the Low Countries, and the monasteries. At times they refer to their sources explicitly, yet one could spend many years of scholarly research tracing the affiliation of their ideas, as well as proving their influence on latter-day works. But on the whole it is the impact of the major problems of their day that forms the content of these works.

---

14 A few examples would be: Emilio Bertana's book on Doni, *Un socialista del cinquecento,* Genoa, 1892; Benoît Malon, *Histoire du socialisme* (includes Doni, Bonifacio, Campanella), Milan, 1879; Karl Kautsky *et al., Die Vorläufer des Neueren Sozialismus,* Stuttgart, 1895; and K. Kautsky, *Thomas More and his Utopia,* New York, 1959 (first published 1887).

Paradoxically, while bestowing respectability on the utopias, it was the Marxists who also made the term 'utopian' into a derogatory phrase by distinguishing between their own 'scientific socialism' and the 'utopian socialism' of all their predecessors.

The great geographical discoveries undoubtedly inspired some of our utopists: they created the opportunity to locate the perfect society somewhere in the far distance to be discovered by imaginary travellers. They also constituted a proof that the present age was surpassing the age of the classics and therefore humanity was not doomed to degeneration. And they may have also supplied the imagination with details of dress, food, customs. But all these factors do not support a claim, made by many, that all Renaissance utopias are based on descriptions and observations of travellers.[15] It is really only a minority among them that used the New World even for the fictional framework. Most of them were satisfied to build their imaginary cities on the soil of Europe or not to give it any locus at all. It is quite striking, in fact, how little influence the New World had on the content of utopias during the Renaissance; one has the impression that they could all have been written even if Columbus had never set sail. The utopists grappled with the problems that beset the civilization of Europe, yet they had no intention of sacrificing that civilization for the sake of the 'primitive', alien cultures in the newly-discovered lands. Therefore, to attempt to prove that More was not fabricating a new society but merely repeating the tales of travellers about the (yet undiscovered) land of the Incas,[16] or to say that the utopists invented their happy lands simply because their admiration was aroused by the social organization of the Indians or by the customs of the Chinese, is, to my mind, to misinterpret the intentions of these writers, and to confuse inspiration with influence. The utopias are, I believe, but additional evidence of the little real interest Europeans had in the human aspects of the 'El Dorado' that had been discovered, and the extent to which they saw it through the spectacles of the European experience.[17] This is all the more obvious in the pathetic attempts to realize the utopias by using the Indians of the New World as the human material.

The utopias of the Renaissance are all visualized as cities. Even when it is an entire country that is ruled by the laws of the ideal state, the author is satisfied with describing just one town, the capital, because 'The person who knows one of the cities will know them all.'[18] And rarely are more than a few lines devoted to life in the countryside. Many reasons can be found for that. 'The first utopias we know were fabricated in

---

[15] S. B. Liljegren, *Studies on the Origin and Early Tradition of English Utopian Fiction*, Uppsala, 1961; D. W. Thompson, 'Japan and the New Atlantis', *Studies in Philology*, Vol. XXX, 1939; and many others.

[16] Arthur E. Morgan, *Nowhere Was Somewhere*, Chapel Hill, 1946.

[17] J. H. Elliott, *The Old World and the New 1492-1650*, CUP, 1970.

[18] Thomas More, *Utopia*, Yale University Press, 1967, p. 63.

Greece and . . . the Greeks were never able to conceive of a human commonwealth except in a concrete form of city. . . . Once this tradition was established, later writers, beginning with Thomas More, found it easy to follow, all the more so because it had the advantage of mirroring the complexities of society within a frame that respected the human scale.'[19] And also because 'la ville manifeste le Règne de l'homme',[20] while the countryside is much more dependent on the vagaries of nature. Finally, because the utopias belong to the Renaissance fashion of presenting models, almost in the sense of Platonic ideas. Together with the model courtier and the model prince, and the model Christian knight, we are presented with a model city which is the smallest unit to contain an entire society.

P. O. Kristeller wrote that he had been 'unable to discover in the humanist literature any common philosophical doctrine except a belief in the value of man and the humanities and in the revival of ancient learning.'[21] This statement adequately applies to the utopian section of Renaissance literature. The variety is such that any generalization is practically impossible. One way out of the maze is classification. Our utopias could be classified in at least four different significant ways: geographically, chronologically, according to the religious denomination of the author, or according to his social status. But one soon learns that these groupings hardly shed any new light on the content of the texts, or explain similarities and differences. As we shall see, there is sometimes more of a resemblance between two utopists from different countries, a few generations apart, and adherents of separate churches, than between two countrymen of the same generation and religious denomination. It seems, therefore, that classification will merely add to the confusion. Another way out is to grade the utopias on two different scales: one, from the most radical to the most reactionary, the other from the most idealistic to the most realistic. The radical–reactionary scale is dependent to a large extent on the personal views of the individual analysing the utopias and on the time in which he is writing, because what appears to him as revolutionary was not necessarily regarded as such by the public in the sixteenth century. The second scale is more relevant to the purpose of this study, because the more realistic the utopia — i.e. the more it contains all the components and necessities of

---

[19] Lewis Mumford, 'Utopia, The City and the Machine', in F. E. Manuel (ed.), *Utopias and Utopian Thought*, London, 1973, p. 3.

[20] Ruyer, op. cit., p. 43

[21] Paul Oskar Kristeller, *Renaissance Thought: The Classic, Scholastic, and Humanist Strains*, New York, 1961, p. 22.

community life — the more interesting it is às a reflection of the true, substantial problems of the age. Rabelais's 'Abbey' is no doubt a delightful flight of fancy, but it is by far the most idealistic utopia since it completely ignores the questions of production, labour, government. Thus it is closer to the 'Land of Cokaygne' type of escapist dream than any of the other genuine utopias.

It is therefore the more realistic utopias that will supply us with most of the information related to the four major issues which will be discussed in this study: Health, Education, Work and Welfare, Law. These will include all the *social* aspects of an ideal community life such as the structure of the family, the status of women, town-planning, class divisions. On the other hand, subjects such as politics, religion, philosophy, unless directly affecting the social questions, will be excluded. And not only because of lack of space. The utopias of the Renaissance (or at least many of them) have been discussed extensively as a chapter in the history of ideas — somewhere in the middle of the road from Plato's *Republic* to B. F. Skinner's *Walden Two*;[22] as parts of studies done on individual authors; as well as tracts in philosophy and in religion of the period. But little has been done so far in examining them as a group of social programmes that are products of contemporary conditions.[23]

This study is confined to Western Europe during the later Renaissance. It has no claim to be an exhaustive analysis of all Renaissance utopias. There may be several others hidden in libraries and archives which have never attracted the attention of publishers and scholars. But from our limited selection it appears that Italy was richest in utopian thought. In addition to the general arguments which explain why Renaissance culture began and spread from this particular corner of Europe, one could claim that the diversity of political systems existing side by side and changing not unfrequently, had led to speculations on the best polity and to an awareness that no single system was divinely ordained. Moreover, the administrative skills exhibited by the Italian city-states and acclaimed by many, may have served as an inspiring example of what a central authority was capable of achieving by sound administration. A negative question, however, would be more difficult to answer. Why were there some countries or nationalities which (to the

---

[22] The latest addition to the histories of utopian thought is F. E. Manuel and F. P. Manuel, *Utopian Thought in the Western World,* Harvard University Press, 1979.

[23] A work which had applied the same method, though only to one or two of our utopias among different writings of the same period, is the book by Helen C. White, *Social Criticism in Popular Religious Literature of the Sixteenth Century,* New York, 1944; J. C. Davis, *Utopia and the Ideal Society,*CUP, 1981, appeared while this present work was already in press.

best of my knowledge) did not produce in that period even a single utopia proper? Claude Backvis attempted to give an explanation for Poland. According to him the Polish thinkers did not have to resort to the 'desperate' utopian method since their country was going through such great upheavals and transformations that there were sufficient reasons to believe that serious political programmes and plans for reform would be heeded and realized.[24] But why none in Spain?[25] Was it the preoccupation with colonization in the New World which diverted the attention of the Spanish humanists from imaginary societies to real social experiments? Or was it the rule of the Inquisition and the enforced religious uniformity that prevented the appearance of utopias proper which are, as was indicated above, a product of a fundamentally secular impulse: to create a perfect society here and now by human powers alone? These are merely tentative speculations which require further research beyond the scope of this study. Here we shall focus on the existing utopias and on their social content, after a brief introduction to each of the utopists.

B. THE UTOPISTS AND THEIR WORKS

It would be impossible to treat all the utopists equally in a brief presentation since some are well known while others had received very scant attention. To the famous I shall grant only a few sentences to put them in the context of the present analysis; the less known will receive a fuller introduction to their background, a short sketch of the content of their utopias, especially on matters which will not be discussed extensively in the following chapters, and references to the most relevant works of scholarship in which they are discussed. The order is chronological according to the date of composition of the utopia.

Thomas More
(1477-1535)

There are innumerable interpretations that trace the sources from which More allegedly derived ideas for his *Utopia*. Their cumulative effect not

---

[24] Claude Backvis, 'Le Courant Utopique dans la Pologne de la Renaissance', in *Les Utopies à la Renaissance*, Brussels, 1963, p. 165.

[25] In 1975 an anonymous manuscript of a Spanish utopia was discovered and published by Stelio Cro (ed.), *Descripción de la Sinapia, peninsula en la tierra austral. A Classical Utopia of Spain*, Hamilton, McMaster University. But *Sinapia* was written around 1682 and no Spanish utopias are known before that date. See also Stelio Cro, 'The New World in Spanish Utopianism', *Alternative Futures*, Summer 1979, Vol. 2 no. 3 pp. 39-53.

only emphasizes the man's erudition, but even more the originality and uniqueness of the mind that could integrate them all into a cohesive whole. From our point of view, however, his literary background is of secondary importance, as the purpose of this study is to investigate reactions to contemporary problems.

More's impact on succeeding utopias is taken for granted. Obviously, he inspired others to make use of the utopian vehicle for expressing social criticism and to adopt some of his ideas for the reconstruction of society. But more important still is the fact that he had instigated a new approach to social problems, which was secular but based on ethical considerations, 'scientific' yet full of compassion for human misery, linking the minutiae to the major issues of state organization. His choice of the utopian form, together with his wit and literary gifts, gained for his work an unsurpassed popularity and influence throughout Europe, which no dry political tract could ever achieve. Some of the succeeding imaginary societies may appear at times as pale reflections of More's *Utopia*, a tribute to his preeminence. Yet each of the following utopias acquired a distinct character through the temperament of its author, his encounter with a different environment and a dissimilar set of beliefs and ideals. Thus, in a way, *Utopia* constitutes a baseline for our comparative study.[26]

Johan Eberlin von Günzburg
(c.1460-1533)

Eberlin's *Wolfaria* does not appear in any of the anthologies or surveys of utopias,[27] but it definitely deserves a place among the imaginary ideal states as the only example of a comprehensive vision of social perfection to come out of the circle of Luther's early followers.

Johan Eberlin was born in a small village near Günzburg in Swabia. He studied theology at the universities of Ingolstadt, Basle, and Freiburg im Breisgau. Then he became a Franciscan observant friar and served as a preacher in the area of Tübingen and Ulm. He broke with the order in 1521 and joined the ranks of those propagating Luther's New

---

[26] The best modern edition of *Utopia* appears as Vol. 4 of *The Complete Works of St. Thomas More*, edited by J. H. Hexter and Edward Surtz, Yale University Press, 1965. It contains the Latin original with a page by page English translation. The English translation was also published separately as St. Thomas More, *Utopia*, edited with introduction and notes by Edward Surtz, Yale University Press, 1967. This is the edition quoted in the present study.

[27] Its significance as a utopia was pointed out by Susan Groag Bell, 'Johan Eberlin von Günzburg's *Wolfaria* — The First Protestant Utopia', *Church History*, Vol. XXXVI, no. 2, June 1967. The biographical data on Eberlin are based on her article. See also L. W. Spitz, 'Johannes Eberlin', *New Catholic Encyclopedia*, 5: 28-9.

Evangelism. At the same time he began to publish numerous pamphlets promulgating the new cause. His utopia is one of these early works and it consists of pamphlets nos. 10 and 11 in the series *Fünfzehn Bundsgenossen,* published in 1521.[28]

Without a tale of an imaginary voyage or a description of the place and the life of its inhabitants, Eberlin presents the new statutes adopted by the elders of an imaginary state which had been suffering from exactly the same evils as Germany. The first section — 'New statutes which Psitacus[29] has brought from the land of Wolfaria, concerning the reformation of the Spiritual Estate' — deals with the status of priests as teachers of the Gospel, the abolition of the monastic orders, rites, holidays, sacraments, and marriage. All of these laws reflect Luther's early teachings, except that on some points Eberlin is more violent (especially in the methods of exterminating his former colleagues, the mendicant friars); sometimes he goes further than his mentor (as in his insistence on divorce on grounds of incompatibility), and on certain points, where Luther's ideas had not yet been clarified, he differs from the final doctrines of the Lutheran Church (as in the number of sacraments). This section of Eberlin's programme differs from other utopias in that it offers not only a description of the ideal state of affairs, but also of the methods, both drastic and gradual, by which the reformed society abolishes the old order.

The second part — 'Description of a new order of the Secular Estate of Wolfaria as reported by Psitacus' — includes the laws concerning all other aspects of community life in a state governed by elected salaried officials who are given titles of nobility.

Wolfaria is radically different from More's *Utopia*: not only does it lack the literary merits, the wit and charm of the latter, but it was also written for a different public and with a different purpose. *Wolfaria* is not a scholarly amusement for the international intelligentsia, but an attack, full of venom and violence, intended to incite the German people. It is written in the vernacular, it echoes the apocalyptic notes of the *Book of A Hundred Chapters,* and expresses the grievances of the

---

[28] The text is to be found in Johan Eberlin von Günzburg, *Ausgewählte Schriften,* Band I, Halle, 1896, pp. 107-31; in *Neudrucke deutscher Litteraturwerke des XVI und XVII Jahrhunderds,* nos. 139-41. Hereafter it will be cited as 'Eberlin'. (The passages quoted from this work in the following chapters were translated by me, as were all other quotations from works which had not been translated previously into English.)

[29] Psitacus was the latinized name of Eberlin's cousin, Ulrich Sittich, which he used often in his writings. Psitacus has no role in the utopia apart from being the person who supposedly had brought these invented statutes to the knowledge of Eberlin.

peasants and the poor artisans. (As S. G. Bell points out, *Wolfaria* includes six out of the Twelve Articles of Memmingen which were circulated in 1524).[30] And yet this utopia is not merely a pamphlet of revolutionary zeal: it contains wider conceptions as regards the responsibilities of the state in matters of trade, education, health, and welfare, in which Eberlin's humanistic training comes to light.

Eberlin spent the year 1522 at Wittenberg in the company of Luther and Melanchton. His later publications are milder than *Wolfaria*. In 1526, for example, Eberlin joined Luther's call to the peasants to refrain from attempting to overthrow the existing order in a pamphlet titled 'True Warning to the Christians in the Margravate of Burgau, that they should guard themselves against Revolt and False Preachers'.[31] In 1525 he became evangelical pastor to Count George II of Wertheim am Main, but was dismissed in 1530 when his patron died. He then procured a living at the village of Leutershausen (near Ansbach) where he spent the last years of his life.

Johan Eberlin does not rank among the major Reformers or the major Humanists; his utopia had no apparent influence on succeeding works of the same genre, not even on those written by his countrymen such as Stiblinus or Andreae. But his vision of a perfect society is an authentic reflection of the sense of new horizons opening to the German people in the early 1520s, before the suppression of the revolts of the knights and the peasants, and when Luther's teachings were still a banner in the fight for a new social order.

François Rabelais
(1483?-*c*.1533)

It is not easy to justify the inclusion of 'The Abbey of Thélème'[32] in an analysis of social utopias, for it is decidedly not a comprehensive description of an entire society. Moreover, even for the inhabitants of this monastery in reverse, it is only an abode for a few years of their youth, before they marry and emerge into the world. Rabelais made no attempt to offer remedies to the evils he saw or a blueprint for a model

---

[30] S. G. Bell, op. cit., p. 12.

[31] See Kyle C. Sessions, 'Christian Humanism and Freedom of a Christian: Johan Eberlin von Günzburg to the Peasants', in L. P. Buck and J. W. Zophy (eds.), *The Social History of the Reformation*, Columbus, Ohio, 1972, pp. 137-55.

[32] 'L'Abbaye de Thélémites' consists of chapters 52-7 in the book of *Gargantua* which was first published in 1534, two years after the book of *Pantagruel*. The first complete edition of the story of *Gargantua et Pantagruel* appeared only in 1564. The edition quoted in this study is the modern English translation by J. M. Cohen, in Penguin Classics, 1955.

society. Nevertheless, his feelings as to what is wrong with the world are expressed by what he explicitly excludes from his imaginary haven.

And Rabelais was a man acquainted with all walks of life: as a son of a rich landowner and a lawyer, a monk of both the Franciscan and the Benedictine orders, an ordained priest, a student of medicine at several French universities, a Greek scholar, secretary and doctor to bishops and cardinals, a physician in the hospital of Lyons, he was unfamiliar with hardly any aspect of life in France. All his rich and varied experience comes to life in the kaleidoscopic book of *Gargantua and Pantagruel,* while in its utopian chapters he offers a refuge from all that he had come to despise in the world. More than any other of our utopias, Rabelais's Abbey is an escapist dream; but for this very same reason it is a more open admission of the author's innermost wishes.

There are numerous allusions to More's *Utopia* in Rabelais's masterpiece, particularly in his choice of names for imaginary places and people. But, as Saulnier has pointed out, these are concentrated mainly in the first book of *Pantagruel* (published in 1532), whereas in the later parts he seems to have forgotten the existence of Utopia.[33] In any case, the 'Abbey of Thélème' bears no relation to More's imaginary state; if at all, it can be seen as an unintentional satire on all the extremely regimented utopian states.

Francesco Patrizi da Cherso
(1529-1597)

Patrizi was born in Dalmatia, studied at Ingolstadt, Padua, and Venice, travelled widely, and served as secretary to several Venetian noblemen. In 1578 he was appointed professor of Platonic philosophy at Ferrara, and in 1592 he was called by Pope Clement VIII to lecture on Plato at the university of Rome. His major work is *Nova de universis philosophia* (1591), which deals with the properties of light, the hierarchy of being, the theory of the soul and the physical world. His other numerous writings show a great variety of interests: Italian poetics, rhetoric, history, and geometry. He translated treatises attributed to Zoroaster and Hermes and wrote on Pythagorean numbers. He tried to prove that their teachings and Platonic philosophy agree with the Catholic doctrines, whereas Aristotle is in conflict with the Church. In the history of science his importance rests primarily on his original views

---

[33] V. L. Saulnier, 'Mythologies Pantagrueliques. L'Utopie en France: Morus et Rabelais', in *Les Utopies à la Renaissance,* Brussels, 1963, pp. 137-62.

concerning space. P. O. Kristeller writes: 'there are good reasons for grouping Patrizi, along with Telesio and a number of other Italian and European thinkers of the sixteenth century, among the Renaissance philosophers of nature, who were unattached to the classic traditions of Western thought and prepared the way for the new science and new philosophy of the seventeenth century and modern times.'[34]

La Città Felice,[35] written probably in 1551 and published in 1553, together with discourses on honour and on poetic madness, is one of Patrizi's earliest works. The influence of Plato's *Republic* is unmistakable both in the form — philosophical speculation rather than an imaginary voyage — and in the content which is concerned solely with the ruling classes and their way of attaining some vague state of happiness and wisdom (the 'celestial waters'). There is also an oblique expression of admiration for the Republic of Venice with its rigid rule of the aristocracy. Patrizi's utopia differs from other Renaissance visions of social perfection in its total rejection of the rebellion against social injustice.

Anton Francesco Doni
(1513-1574)

Doni was born to a poor family in Florence. Out of necessity rather than conviction he joined a monastic order and became a priest. But in 1540 he relinquished the monastery and began a life of restless travel and varied activities. He studied law at the university of Piacenza; he established his own printing-press in Florence, but then moved to Venice to work for another printer; in Florence he had been a secretary to the Florentine Academy sponsored by the Medici, in Venice he helped to found another literary group — the Accademia Pellegrina. He soon became one of the *poligrafi* — those prolific writers who wrote popular literature in the vernacular for the Venetian presses. His lifelong interest was music, but he wrote on almost every subject and in every literary form.

Doni's utopia is the sixth of the seven worlds described in *I Mondi* (1552).[36] The fictional framework is a dream unravelled in a dialogue

---

[34] Paul Oskar Kristeller, *Eight Philosophers of the Italian Renaissance*, London, 1965, p. 110.

[35] *La città felice* (Venice, 1553) was reprinted in Carlo Curcio (ed.), *Utopisti e riformatori sociali del cinquecento*, Bologna, 1941. This is the edition referred to in the present study.

[36] 'Il mondo savio e pazzo' in Carlo Curcio (ed.), *Utopisti e riformatori*, pp. 3-15, is the edition referred to in this study.

between Pazzo (the fool, the madman) and Savio (the wise). The use of Pazzia (folly) to eulogize all that is simple and sensible is taken directly from Erasmus.[37] Other influences are those of More (Doni had edited Ortensio Lando's Italian translation of *Utopia* and added to it an introductory letter of praise (1548)), and of Guevara (in *I Marmi* (1553) Doni expressed approval of the wise laws of the Garamanti, the quasi-utopian people in *Relox de los Principes*). But Doni's outlook was profoundly different from that of the other utopists. He belonged, together with Ortensio Lando and Nicolò Franco, to a small group of Cinquecento social critics who, in their exasperation with the injustice of the social order, rejected also most of the values of Renaissance civilization. They discarded the belief in education and learning as a path to a better world; they put no trust in rulers, however enlightened; they had no alternatives to offer but the total destruction of the existing order. As P. F. Grendler writes: 'They conveyed disorientation, disillusionment, and pessimism for their generation in the difficult years between 1530 and 1560.'[38]

Thus, Doni's 'Wise and Mad World' is a fantasy, ranking low on the scale of realistic imaginary societies, far more drastic in its repudiation of all existing institutions than even More's or Campanella's utopias, an expression of acute desperation.

In the nineteenth century Doni's utopia aroused the interest of socialists who attempted to depict him as a precursor of Marx because of his attack on the maladministration of property and the exploitation of the toiling masses by the idle rich. Other historians paid greater attention to his poetry, philosophy, and activities as printer and publisher.[39] Only in the last three decades, thanks to the works of Curcio,[40] Firpo,[41] and Grendler, Doni's utopia has gained prominence as the conclusion and concretion of the ideological premises of the thought of Doni himself and of his circle of social critics.

---

[37] The same motif is the theme of the anonymous quasi-utopia, *La Pazzia* dating from the same period as Doni's imaginary society.

[38] Paul Grendler, *Critics of the Italian World 1530-1560: Anton Francesco Doni, Nicolò Franco and Ortensio Lando,* University of Wisconsin Press, 1969, p.14. See also Grendler's 'Utopia in Renaissance Italy: Doni's New World', *JHI,* Vol. 26, 1965.

[39] A comprehensive bibliography of Doni's writings and of all works on him was compiled by Cecilia Ricottini Marsili-Libelli. *Anton Francesco Doni: scrittore e stampatore,* Florence, 1960.

[40] Curcio, *Utopisti e riformatori,* as well as his *Dal Rinascimento alla Controriforma,* Rome, 1934.

[41] Luigi Firpo, *Il pensiero politico del Rinascimento e della Controriforma,* Milan, 1966.

Gasparus Stiblinus (or Kaspar Stüblin)
(1526-1562)

De Eudaemonensium Republica Commentariolus[42] is the creation of a German humanist scholar. Like Johan Eberlin von Günzburg, Stiblinus too was a Swabian, of a poor family, who attended the university at Freiburg im Breisgau — yet there the similarity ends. Stiblinus was a Catholic. He wished to see the Church purified from its abuses and prayed for a reunification of Christendom on the basis of Erasmian reforms and the conciliatory proposals of the Peace of Augsburg. *Coropaedia*, the first book in the volume that contains the utopia, is a proposal for a reformed *regula* of a convent. It includes an attack on the vices of contemporary monastic life, and calls on the novices and nuns to adhere to the utmost simplicity, modesty, and devotion.

Stiblinus was a humanist only in the strict sense of the word, i.e. a believer in classical learning. He was a teacher of Latin and Greek at the schools and university of Freiburg im Breisgau, in the schools of Schlettstadt in Alsace, and for a short period he had the chair of Greek in the newly founded university of Würzburg. His major life-work was a translation of the tragedies of Euripides into iambic Latin verse (1562).

Stiblinus's attitude to society was that of a schoolmaster towards a group of unruly, wicked children who must be constantly watched, reprimanded, and instructed. The inhabitants of his imaginary island are ruled by a small minority of virtuous and learned aristocrats, and lead a life of sobriety, simplicity, and piety. He exhibits little interest in social and economic matters. The didactic moralizing tone is uninspiring, the created atmosphere is harsh and grim, the content belies the names of happiness.[43] Yet this utopia by an obscure professor is probably more typical of prevalent attitudes to the social order than are the elaborate creations of the renowned utopists who were, after all, outstanding and exceptional thinkers.[44]

---

[42] Kaspar Stiblin, *De Eudaemonensium Republica,* con una introduzione e la bibliografia dell'autore a cura di Luigi Firpo, Turin, 1969. This book contains a facsimile of the original edition published in Basle in 1555. Firpo's introduction gives detailed information about Stiblinus's life, work, and the fate of his writings. In a French translation Firpo's introduction appears also in *Les Utopies à la Renaissance.*

[43] The imaginary island is named Macaria ('the land of the happy'). It was first used by Thomas More for a country neighbouring Utopia. The capital of Stiblinus's Macaria is Eudaemon (referring again to the happy or the blessed).

[44] Stiblinus and his utopia were first mentioned in an English work in W. Begley, *Nova Solyma: An Anonymous Romance* (London, 1902). The inaccuracies of Begley are repeated in more recent works of scholarship which refer to Stiblinus.

Ludovico Agostini
(1536-1612)

'The ideal state of the Counter-Reformation', 'the first post-tridentine utopia' are the titles awarded by Firpo[45] to Agostini's imaginary republic. Grendler describes it as 'a Catholic version of Calvinist Geneva superimposed on the typical Renaissance utopia'.[46] Both put the emphasis on the religious aspects, the ascetic spirit, and the two parallel hierarchies of clerical and secular administration. In the context of this study, however, Agostini's utopia is of particular interest because it is one of the most 'realistic' imaginary states. It is not based on radical formulas of change but on the improvement of the existing system, and it deals with the entire spectrum of social life thus reflecting both the multifariousness of contemporary social issues and the solutions for feasible reforms.

Agostini was a native of Pesaro and studied law at the universities of Padua and Bologna. After a brief military adventure he returned to Pesaro and served his town as a lawyer and administrator. He was by nature profoundly religious and inclined to melancholy and pessimism. Thus, in 1582 he retired from the world of affairs to Soria, a small village north of Pesaro, where he divided his time between music and poetry, religious meditation, and work in the fields. He only emerged from his retreat to go on pilgrimages, including a long voyage to the Holy Land in 1584-5.[47]

There is no work by Agostini entitled 'Repubblica Immaginaria'; this was the title adopted by modern Italian scholars for a number of chapters in Book II, part II, of Agostini's dialogues of *L'infinito,* which were composed in the years between 1585 and 1590.[48] He arrives at the description of an ideal state from a discussion on the origin of laws, carried out between two symbolic interlocutors: Infinito, who personifies divine wisdom, and Finito who represents human knowledge based on historical experience. The text is written in pedantic prose, laden with citations and digressions. The reflection of the ideals of the Counter-Reformation is, indeed, unmistakable; but the acute social malaise, which motivated Agostini to present his remedies in utopian form, was

[45] Luigi Firpo, *Lo stato ideale della Controriforma: Ludovico Agostini,* Bari, 1957.

[46] P. F. Grendler, *Critics of the Italian World,* p. 166.

[47] L. Agostini, *Viaggio di Terrasanta,* 1585.

[48] Luigi Manicardi, 'La Repubblica Immaginaria di L. Agostini', *La Rassegna,* Genova, XXXIV, 1926. Carlo Curcio reprinted only two-thirds of the relevant text in *Utopisti e Riformatori;* the entire utopia was first published as Ludovico Agostini, *La Repubblica Immaginaria,* edited with an introduction by Luigi Firpo, Turin, 1957.

primarily a consequence of the conditions in the towns of Italy. In this he resembles all other utopists, irrespective of their religious denomination.

Tommaso Campanella
(1568-1639)

La Città del Sole is the only utopia to have been written inside a prison cell. It is the creation of one of the most extraordinary and tragic figures of the late Renaissance and, not without cause, it ranks with More's *Utopia* for the interest it aroused among scholars and social thinkers ever since the seventeenth century.[49]

Campanella was a son of illiterate parents from Calabria, a Dominican monk, a disciple of Bernardino Telesio and a defender of Galileo, the leader of the Calabrian insurrection against the Spaniards, all his life a victim of persecution by the Inquisition, a protégé of Richelieu, and the author of over 100 works ranging from lyric poetry to philosophy, written mostly within the walls of prisons where he spent over thirty years of his life under sentence for heresy and sedition.

In the odd eclectic fusion of rationalism and mysticism, astrology and experimental philosophy, messianism and nationalism, heresy and staunch orthodoxy, the recurring *leit-motif* of Campanella's work is that of radical reform: reform of the sciences from useless classical erudition to a philosophy based on experience: a theologial reformation, founded on natural religion, which would reconcile all faiths; a vision of a universal monarchy (headed by the Papacy, or Spain, or France) that will unite the world. His *Monarchia di Spagna* (1599) and *Monarchia del Messia* (1605) are also utopias, but of universal peace in a future golden era, inspired by the same millennial dream as the revolt in Calabria and the *City of the Sun*. The latter was regarded by Campanella himself as the embodiment of all his ideals, and for this reason he kept revising it throughout his life. It was not a fantasy, but a programme for action, the first step towards a total reformation of the world. All his beliefs in astrology, experimental science, education, metaphysics based on a trinitarian structure of power, wisdom and love,[50] natural religion, the supremacy of the community

---

[49] Campanella rewrote his utopia several times: the original text was written apparently in 1602 in Italian, and the second vernacular version in 1611; then there are two Latin versions, of 1614 and 1632. There are good reasons to suppose that it began to circulate in manuscript as early as 1607, but was first published, in Latin, in 1623. The best modern edition is Tommaso Campanella, *La Città del Sole*, edited and annotated by Norberto Bobbio (Turin, 1941) — it includes both the Italian and Latin texts with footnotes indicating the differences between the various versions.

[50] Compare his *Metafisica* (1638) with the symbolic rulers of the City of the Sun: Metaphysicus (Hoh), Pon, Sin, and Mor.

over the individual, and the urgent need to alleviate human misery by a communal ownership of goods, are incorporated in his utopia.

It should be emphasized that the religion of the Solarians is not that of a primitive society which awaits a revelation to become Christian; in the City of the Sun, Christ and the apostles are known and revered among the great lawgivers in history. But Campanella, unlike More, presents this non-Christian faith as the rational belief of wise men who are acquainted with all churches yet adhere to none. The Christian apologia which tries to explain away the pagan religion in More's *Utopia*, is by no means applicable to Campanella's utopia.

Beside his impressions of life in the Italian world and the experience of monastic life, Campanella's sources of influence are mostly literary: Plato and More, Iambulus' 'Islands of the Sun' as reported by Diodorus Siculus, Pythagorean philosophy, the myth of Sparta, Botero's accounts of the customs of the Chinese, and perhaps also tales of the Aztec and Inca civilizations. This first utopia of the seventeenth century opens a new era of utopian writing which is to include Andreae and Bacon; but it also has a unique position all of its own owing to the extraordinary qualities and beliefs of its author.[51]

I. D. M. Gentilhomme Tourangeau

The author of *Histoire du Grand et Admirable Royaume d'Antangil* (Saumur, 1616) remains unidentified.[52] The available information about him is based solely on the text of his utopia: a nobleman from Touraine; most probably an army officer (there are not less than fifteen chapters on military affairs, and in the prefatory epistle he excuses his literary faults by saying: 'ceux qui font profession des armes ne sont pour l'ordinaire si délicats et excellens escrivaines');[53] and obviously a Protestant, though neither a Calvinist nor a sectarian. It is also clear that he had read More's *Utopia*.

---

[51] There is no full English translation of the *City of the Sun*. Henry Morley (ed.), *Ideal Commonwealths*, London, 1886, contains a translation by R. W. Halliday, but all sections on sexual matters were expurgated. In G. Negley and J. Max Patrick, *The Quest for Utopia*, New York, 1952, a translation by J. Gilstrap omits passages on astrology and military affairs. Wherever possible I quote one or the other of the English translations, but when necessary I refer to the Italian text in the edition of Bobbio.

[52] This utopia was first mentioned by Frédéric Lachèvre (ed.), *Les Successeurs de Cyrano de Bergerac* (Paris, 1922). An attempt to identify the author was made by M. Nicolas Van Wijngaarden, *Les Odysseés philosophiques en France entre 1616-1789*; but his theory was refuted by F. Lachèvre in his introduction to the text of the utopia which he edited and reprinted as *La Première Utopie Française: Le Royaume d'Antangil*, Paris, 1933. This is the edition referred to in this study as *Antangil*. Summaries of *Antangil* in English can be found in G. Negley and J. Max Patrick, *The Quest for Utopia*, and in F. E. and F. P. Manuel, *French Utopias: An Anthology of Ideal Societies*, New York, 1966.

[53] *Antangil*, p. 25.

The story begins with the customary imaginary voyage, but the author does not reach the shores of the ideal state — he receives a full account of the kingdom of Antangil from its ambassador in Java. The long and detailed narrative includes a description of the Kingdom and its beautiful provinces, the government, the army and police, the education of its youth, and its religion. Although there are certain similarities to More's *Utopia*,[54] it is a fundamentally different vision of social perfection: equality and communism are replaced by a rigid separation of the classes, and the frugality common to most utopias is replaced by splendour, ceremony, and display. The religion of Antangil bears close resemblance to the Church of England: there is an ecclesiastical hierarchy from curates to bishops, but it is not subordinated to a supreme head elsewhere, nor does it have any legal prerogatives; the clergy dress in impressive costumes, but celibacy is not required of them; they believe in good works as a sign of faith, but have no images in their churches; there are two sacraments — baptism and holy communion; prayers to the saints or for the dead are prohibited and purgatory denied.

His version of Protestantism could be the key to his inspiration for the entire work: it seems to be an idealized conception of England, with a constitutional monarchy, a senate of two chambers — one of representatives from all the provinces, the other of the greatest in the land, and an upper class which sends its sons to the Academy to be trained for the responsibilities in government.

Johann Valentin Andreae
(1586-1654)

Andreae's utopia begins in an allegorical fashion: a voyage in the good ship Phantasy upon the Academic sea, an escape from tyranny, sophistry, and hypocrisy, and a search for knowledge and piety. The pietist spirit and the mystical aspects, which include alchemy and angelology, give it an appearance which is more foreign to the modern reader than the practical, down-to-earth atmosphere of More's *Utopia*. Yet underneath that plumage one finds a carefully thought-out plan for the organization of a small city-state.

The *Reipublicae Christianopolitanae Descriptio* (1619)[55] received admiring recognition from several intellectuals during the seventeenth

---

[54] Compared by Lachèvre in the introduction to *Antangil*, pp. 13-20.

[55] Felix Emil Held, *Johann Valentin Andreae's Christianapolis: An Ideal State of the Seventeenth Century*, Oxford University press, 1916, contains a complete English translation and a long introduction. The utopia will be referred to as 'Andreae', the introduction as 'Held'.

century and influenced the thought of Comenius, Hartlib, Bacon, Samuel Gott, and the founders of the Royal Society. Later it aroused the interest of philosophers like Leibniz and Herder.[56] However, it never attained the fame of More's and Campanella's utopias, perhaps because it did not appeal as much to the socialists of the nineteenth century.

Andreae, a native of Württemberg, a son of a Lutheran theologian, studied six years at the university of Tübingen and acquired a wide education in languages, mathematics, history and literature. After several years of travel, he served as a teacher and preacher at Vaihingen, Calw, and Stuttgart. In Calw he founded a mutual protective association of workers. It seems certain that for a number of years he was closely associated with the mysterious Rosicrucian movement.[57] The fact that from 1617 onwards he sometimes referred to them in disparaging terms can be attributed either to caution or to disappointment. In any case, the entrance to Christianopolis is prohibited to the *false* Rosicrucians, not to the true followers of the ideals of renewal in religion, Christian charity and scientific research.

'A Modell of a Christian Society' and 'The Right hand of Christian Love Offered' (1619-20)[58] constitute a programme for the creation of a society based on the same ideals as *Christianopolis,* yet functioning as an organization within existing society rather than as a separate community. Apparently Andreae even made attempts to found such a society, once between 1618 and 1620, and again in Nuremberg in 1628.

Andreae's inspiration for describing an imaginary perfect city came from *Utopia* and the *City of the Sun*; Campanella provided him also with ideas in the sphere of education; the Hermetic-Cabalist sources supplied the esoteric mysticism; Calvin's Geneva was his model for the strict supervision of morals;[59] John Dee contributed the utilitarian approach to mathematics; Paracelsus the advocacy of iatro-chemistry. The crucial element, however, was Andreae's poignant sense of the failure of the

---

[56] On Andreae's influence see Held, *passim;* Frances Yates, *The Rosicrucian Enlightenment,* London, 1972, pp. 125-9, 160-80; Charles Webster, *The Great Instauration,* London, 1975; G. H. Turnbull, *Dury, Hartlib and Comenius,* Liverpool, 1947; H. R. Trevor-Roper, 'Three Foreigners' in *Religion, the Reformation and Social Change,* London, 1972; M. J. Lasky, op. cit., pp. 330-4, 663 nn.14, 15.

[57] There is disagreement among scholars regarding Andreae's link with the Rosicrucians. Frances Yates has argued that, at least for several years, he was the moving spirit in this intangible brotherhood; Charles Webster disagrees, op. cit., p. 249; and so does J. W. Montgomery in his curious panegyric to Andreae: *Cross and Crucible: Johann Valentin Andreae (1586-1654), Phoenix of the Theologians,* The Hague, 1973, p. ix.

[58] These two tracts were found among Hartlib's papers. See G. H. Turnbull, 'Johann Valentin Andreaes Societas Christiana', *Zeitschrift für Deutsch Philologie,* 73 (1954) and 74 (1955).

[59] See his autobiography, *Vita ab ipso conscripta,* p.24 quoted in Berneri, op. cit., p. 105.

Lutheran Reformation — to bring enlightenment, piety, and happiness to European society. As with most of our utopists, Andreae's utopia is the culmination of all his aspirations and his refuge from all that caused him sorrow in this world.

Robert Burton
(1577-1640)

'A Utopia of Mine Own' is a short section in the Preface of Democritus Junior to the *Anatomy of Melancholy*.[60] Burton's masterpiece had enjoyed several periods of vogue among English men of letters, but the utopian section has begun to be considered seriously only in recent times.[61]

The sedate vicar and lifetime Student of Christ Church produced a blueprint for an ideal kingdom which is quite surprising in its exhibition of wide knowledge concerning contemporary economic and social matters, its utilitarian approach to worldly affairs, and concessions to human weakness. It is completely grounded in actuality, despite the fact that his inspiration was derived mostly from his extensive literary erudition.

The author of the *Anatomy* had read the better known utopias of his predecessors — in the version of 1638 he mentions not only Plato and More but also Andreae, Campanella and Bacon — yet he dismisses them as 'witty fictions, but mere *chimeras*', because their revolutionary proposals for complete equality, 'takes away all splendour and magnificence'. Burton's ideas, when not his own, were borrowed from accounts of existing societies rather than imaginary lands. It seems that he was provoked by patriotic shame when he compared the situation of England with that of other nations, particularly the Dutch. His major concern is economic: he attributes the 'melancholy' of England to its restricted manufacture and commerce, the extremes of wealth and poverty, idleness and depopulation caused by enclosures. His programme of cure is based on planning, supervision, encouragement of enterprise, and socialized services. The most 'utopian' aspect is the dependence of the entire scheme on a benevolent dictator who would inaugurate all these reforms. He was, in fact, describing the potential of the processes which were already taking place in England; but he also

---

[60] The edition used for this study was the one published in three volumes by J. Bell and Sons, London, 1926. The utopia is in Vol. I, pp. 109-22.

[61] See J. Max Patrick, 'Robert Burton's Utopianism', *Philological Quarterly*, Vol. XXVII, 1948; William R. Mueller, *The Anatomy of Robert Burton's England*, University of California Press, 1952; Pierre Mesnard, 'L'Utopie de Robert Burton' in *Les Utopies à la Renaissance*.

desired to prevent, or at least to blunt, the injurious effects of these developments on the poorer classes. Thus, although a conservative in matters of church, government, and social hierarchy, his scheme portrays many features of a welfare state.

His utopia is short, expressed in succinct sentences, shorn of any fictional framework or excessive moralizing, and it presents one of the more realistic visions of an ideal state.

Francis Bacon
(1561-1626)

The fame of *New Atlantis*[62] rests on the description of the systematic methods and the achievements of Salomon's House — the institute for scientific research which had inspired the foundation of the Royal Society. But in the context of this study Bacon's utopia has little to offer. The ties between the College and the imaginary society of Bensalem remain undefined: it seems as if the investigation of nature is carried out in isolation and in secret from the rest of society. Hardly any indication is given of the influence of their discoveries on the well-being of the community, or in what way the purported perfection of the social organization advances the research.[63]

Bacon had read Andreae's and Campanella's utopias, but these two far surpass his own in the integration of science in the social framework, in a programme for scientific education, and in the demonstration of the practical use of inventions. Even in purely scientific positions, the earlier utopias (despite their espousal of astrology and alchemy) seem more progressive because of their emphasis on mathematics and their recognition of the new theories of Copernicus, Kepler, and Galileo.

Admittedly, *New Atlantis* is only a fragment, an exercise which Bacon abandoned in order to complete his Natural History. Therefore, his social ideals have to be elicited from the scant information on the history of the island, its methods of maintaining isolation yet benefiting from

[62] *New Atlantis* was written apparently in 1624 and published posthumously, together with the *Sylva Sylvarum* in 1627. The edition quoted in this study is *The Advancement of Learning and New Atlantis*, with an introduction by Thomas Case, OUP, London, 1974.

[63] This point has been made several times. See, for example, Robert P. Adams, 'The Social responsibilities of Science in Utopia, New Atlantis and After', *JHI*, Vol. X, 1949; Judah Bierman, 'Science and Society in the New Atlantis and other Renaissance Utopias', *PMLA*, 78, 1963; Eleanor Dickinson Blodgett, 'Bacon's New Atlantis and Campanella's Civitas Solis: A Study in Relationships', *PMLA*, 46, 1931; J. G. Crowther, *Francis Bacon — the First Statesman of Science*, London, 1960; A. Rupert Hall, 'Science, Technology and Utopia in the 17th Century', in P. Mathias (ed.), *Science and Society 1600-1900*, CUP, 1972.

the achievements of other nations, the Strangers' House, the encounter with the Jew, the laws of marriage, and the Feast of the Family.

Ludovico Zuccolo
(1568-1630)

Little is known about Zuccolo's life. He was born to a noble family in Faenza; for nine years he served the family of Della Rovere as a courtier in Urbino and Pesaro; in 1621 he returned to his home town and later only left it for a short mission to Spain (1623), and for short visits to Venice where his books were published. All his works were devoted to political subjects: the first was *Considerazioni politiche e morali* (1621), which contains a tract on 'Reason of State'.[64] In 1625 he published the *Dialoghi di Lodovico Zuccolo* which include as the ninth dialogue, 'Il Belluzzi, overo la Città Felice' — a panegyric to the republic of San Marino; the fourteenth dialogue, 'L'Aromattario, overo della repubblica d'Utopia" — a criticism of More's *Utopia* for resembling a large monastery rather than a republic; and the thirteenth dialogue, 'Il Porto, overo della repubblica d'Evandria', which is the utopia proper, a description of an imaginary perfect state.[65]

A strong element of Italian patriotism, verging on xenophobia, but explicable in the context of the incessant invasions of foreign armies to Italy during a whole century, is evident in all his writings. For this reason he aroused the interest and respect of his compatriots in later centuries.[66]

In his description of the republic of San Marino[67] he repeatedly praises the wisdom of its inhabitants for retaining their liberty and freedom from foreign influence for over a thousand years. Their secret, he says, lies not only in the natural barriers which protect them, but in their good ancient laws, the absence of extremes in wealth, the simplicity of their life and the careful education of their youth. As we shall see, he founded the imaginary state of Evandria on the very same principles.

---

[64] 'Della Ragione di Stato' in B. Croce e S. Caramella (eds.), *Politici e moralisti del seicento*, Bari, 1930; also in Luigi Firpo, *Il pensiero politico*, pp. 646-71.

[65] These dialogues appear in Ludovico Zuccolo, *La Repubblica d'Evandria e altri dialoghi politici*, a cura di R. de Mattei, Rome, 1944. This is the edition referred to in the present study.

[66] See, for example, Benedetto Croce, *Uomini e cose della vecchia Italia*, Bari, 1943, pp. 194-200.

[67] *Il Belluzzi, overro la città felice* was published separately in Bologna, 1929, with an introduction by Amy A. Bernardy who cites other examples of praise for San Marino, both before and after Zuccolo.

2

LANDS OF HEALTH

Enter not here . . .
. . . you with your sores, gnawed to the bone by the pox,
Take your ulcers elsewhere and show them to others,
Scabby from head to toe and brimful of dishonour,
Grace, Honour, praise, and light
Are here our sole delight,
Of them we make our song.
Our limbs are sound and strong.
This blessing fills us quite,
Grace, honour, praise, and light.

(Rabelais, 'The Abbey of Thélème')

Social justice is irrelevant, collective happiness is inconceivable, in the midst of squalor and disease. The most essential condition for the creation of a perfect community is the cleansing of filth and the banishment of illness. A New Jerusalem cannot be built without an effective sewage system. Utopists of all times, irrespective of their social ideology, presupposed or planned for their ideal lands better health and sanitary conditions than in the existing world. The worse the reality, the more space and thought devoted to these issues in utopias. In the perfect societies described during the Renaissance, the topic of *Sanità* looms very large. If we were to put all our texts together, we should find that this subject probably occupies more space than any other. And although it may seem rather odd at first to associate utopian visions with such down-to-earth issues, one glance at the daily reality of the period suffices to make it clear why solutions to the problems of public health seemed so crucial to contemporary visionaries. The usual associations evoked by the term 'Renaissance', those of art, learning, beauty, and humanism, become all the more remarkable when the foul, noisome, diseased background is revealed.

Much has been written on the great epidemics that had ravaged Europe in the sixteenth and early seventeenth centuries. But it must be remembered that those were but the great catastrophies that punctuated a routine of suffering, innumerable incurable maladies, painful treatments, with the constant accompaniment of accumulated filth and stench. This state of affairs, despite the great advances in the sciences, was not destined to change before the nineteenth century, when the introduction of effective sanitary systems and the revolutionary achievements in practical medicine were to alter the reality of everyday life more perhaps than all political and social revolutions. Our utopists, some of whom are so ardently described as precursors of socialism, were just as much the precursors of public health reformers: Edwin Chadwick paid tribute to Thomas More no less than did Karl Kautsky.

Rabelais simply banished disease by fiat, and to his delightful abbey would be admitted only handsome men and beautiful women without a blemish or an ailment.[1] But all the other writers seriously grappled with the issues of public health. Only rarely did they allow themselves flights of imagination that introduced elements of 'science fiction'. On the whole, no other subject better demonstrates their realistic intentions: they did not dismiss these uninspiring questions, and they endeavoured to offer feasible solutions. Moreover, in no other sphere are the limitations of the age so apparent, since these problems involve not merely the reorganization of institutions — which imposes no fetters on the imagination — but also science and technology. And the utopists were unable to exceed the bounds of the medical knowledge of their day, the limitations of the existing technology, and the prevalent scientific theories and superstitions.

The universal instinct of all men for self-preservation and for freedom from pain and discomfort had always proved more powerful than deterministic supernatural views of disease. The official Christian doctrine, that regarded sickness as a divine punishment for sin, and the astrological view that attributed the outbreak of disease to specific conjunctions of the stars, did not prevent the masses from flocking to healers of all sorts. Nor did they deter the authorities and reformers from waging war on the earthly elements believed to influence human health. In the utopian lands all citizens were supposed to be virtuous, yet the utopists did not claim that therefore, *ipso facto,* they would all be healthy. The good health of the utopians was to be a result of the carefully thought-out battle against disease that the virtuous community would be conducting. Concern for the flesh, tainted with hedonism in the eyes of the Church, was not considered morally wrong by the utopists. On the contrary: some of them declare explicitly that health is the greatest of all pleasures and the basis for all others.[2]

---

[1] Rabelais, pp. 150, 154.
[2] More, p. 99; Patrizi, p. 123.

None of the utopists[3] belonged to the medical profession, and in none of these texts is there a detailed exposition of a medical theory as a basis for the proposed reforms. But the influence of the prevalent theories of the day is quite apparent: the desire to correct the corruption of the air was based on the atmospheric-miasmatic view; quarantine, isolation, and disinfection measures were the result of the growing belief in the contagious nature of certain diseases; the advocated regimen and many of the cures were based on the ancient principles of humoral pathology. One theory did not exclude the others. On the whole, there are no great theoretical differences between the utopists, except for slight shifts in emphasis. The theories may have been false, and therefore some of the prophylactics and remedies may seem utterly futile. Nevertheless, some proposals would be considered quite sound even today: for, whatever the theoretical basis, common sense made it clear then, as it does now, that fatigue, malnutrition or excessive gluttony, evil smells, spoiled food, polluted waters, and lack of medical facilities are all injurious to health.

Different motivations induce men to propose public-health reforms: an altruistic humanitarian wish to alleviate suffering; an aesthetic sense offended by slums, dirt, and ugly diseases; and a religious notion that identifies cleanliness with purity.[4] In the utopias of the Renaissance, as we shall see, all three are evident. In this subject, more perhaps than in any other, the distinction between radical and conservative utopias is totally irrelevant. The plans for a reorganization of public-health administration are at least as advanced in 'reactionary' *Antangil* as in the 'revolutionary' *City of the Sun*. Each and every one of them was aware of the fact that no perfect community could survive in such dire surroundings, and that without proper medical care suffering would remain to mar the ideal. Therefore, each utopia contains a description of the preventive measures and of the medical facilities which the author believed to be necessary for the creation and the preservation of a good society.

A. PREVENTIVE MEDICINE

(1) The City

The problems of public health discussed in the utopias are all urban problems. Reality was at its worst in the big centres, where the density

---

[3] With the exception of Rabelais.

[4] For this notion see, for instance, the cry in Andreae's *Christianopolis* (p. 270): 'Oh, this body of ours! how unclean, how polluted. . . . Pity us, oh Thou source of life, wash and purify that uncleanness.'

of the population was such that all the evils of bad hygiene and sanitation became immeasurably compressed. The rural districts had all the blessings of open spaces, clean water in abundance in streams and ponds, fresh air, easier access to fresh foods. The only advantage of the city over the countryside was the presence of the majority of the medical profession. True, not all cities of Western Europe were exactly the same — but for the purpose of this study, as an impressionistic background for the utopian visions, the differences are negligible.

The level of urban hygiene everywhere was appallingly low, even worse perhaps than in the same — much smaller — towns in Roman times. Public-health administration during the Renaissance was not in any way different from that of the Middle Ages. Municipal authorities made frantic endeavours, especially in crises of epidemics, to enforce sanitary ordinances. The most advanced system of public-health administration was developed in the states of Northern Italy where the Health Boards had been established to fight epidemics in the time of the great pandemic of 1347-51. Initially these offices were provisional, but gradually they became permanent magistracies which, during 'salubrious times', were responsible for all matters pertaining to health such as the marketing of food, the sewage system, the disposal of refuse, the activity of the hospitals and pesthouses, beggars and prostitutes, burials, the medical profession, and the sale of drugs. But even there, the ordinances were obstructed by poverty, ignorance, stupidity, and vested interests.[5] In all European cities similar regulations were repeatedly decreed year after year, but did not succeed in preventing the wholesale pollution of the rivers, the disposal of all refuse into the narrow unpaved alleys, the stench arising from the offal thrown into the gutters by slaughterhouses and fishmongers. Water supplies were permanently deficient and usually polluted; drains were non-existent. Many inhabitants kept livestock in their homes or yards: pigs, geese, ducks, roamed the streets and added their contribution to the general filth. Attempts were made to check the growth of the cities but they only resulted in more overcrowding and deterioration of living conditions. The mean hovels of the slums, made of wood, clay, and straw, vermin-infested and surrounded by refuse heaps, were an ideal abode for the black rat and ideal fuel for the big fires.

Many were deeply concerned about these problems. The College of Physicians in England presented a report in 1630 listing the 'annoyances'

---

[5] Carlo M. Cipolla, *Public Health and the Medical Profession in the Renaissance*, CUP, 1976, pp. 32-5.

that were fostering contagious diseases, citing among them overcrowding, the neglect of cleansing the common sewers and town ditches, stagnant ponds, uncleanness of streets, laystalls too near the city, slaughterhouses in the city, burying of infected people in the churchyards of the city, the selling of musty corn, unsound meat, and tainted fish in the markets.[6] Individuals such as Dr Andrew Boorde and Sir Thomas Elyot clamoured for reforms and remedies. Our utopists joined this campaign in their comprehensive plans for ideal communities.

The utopists, however, had one great advantage over all reformers and authorities. They could roam the world in their imagination in search of an ideal site. Their perfect community was to be built on virgin soil, free from the debris of centuries. Some went very far in their search, to 'the midst of the Temperate Zone, or perhaps under the Equator, that Paradise of the World, *ubi semper virens laurus,* where is a perpetual Spring. . . .'[7] Others were satisfied to remain on familiar ground: 'We shall not travel elsewhere to search for a better site than that in which we find ourselves at present.'[8] But the criteria for the choice of a location were the same: a temperate climate, fresh air, plenty of water resources, fertile soil, and no stagnant swamps. Often they would place their imaginary town on a hill to expose it to salubrious winds.[9] Only Thomas More, wishing to emphasize what can be achieved by human efforts alone, did not grant his Utopians the benefit of a specially chosen location: 'they have not a very fertile soil or a very wholesome climate'. Nevertheless, even he did not begrudge them an ample natural supply of fresh water.[10]

Their second concern, after finding the right place, was the size of the ideal community. The utopists realized that great fluctuations in the size of the urban population were detrimental to control over the sanitary situation and to public health in general. Therefore they endowed the authorities of the imaginary lands with effective power to regulate the size of the city. They implicitly assumed that as a result the population would remain steady. In some utopias (of Doni, Patrizi, Campanella) the state would exercise control over birth rates. Freedom of movement would be restricted: the utopians would not be allowed to travel freely or to settle where they wished. Strangers, though welcomed as guests, would not be easily accepted as permanent residents. The English

---

[6] Quoted in F. P. Wilson, *The Plague in Shakespeare's London,* OUP, 1963, pp. 23-4.

[7] Burton, p. 110.

[8] Agostini, p. 83.

[9] Patrizi, pp. 126-8; Stiblinus, p. 77; *Antangil,* p. 29; Andreae, p. 143; Zuccolo, p. 45; Giovanni Bonifacio, *La Repubblica delle Api,* Venice, 1627, p. 15.

[10] More, pp. 64, 109.

utopists, accustomed to the limitations of an island, showed even deeper awareness of the problem:

But that the city neither be depopulated nor grow beyond measure, provision is made that no household shall have fewer than ten or more than sixteen adults; there are six thousand such households in each city, apart from its surrounding territory. Of children under age, of course, no number can be fixed. This limit is easily observed by transferring those who exceed the number in larger families into those that are under the prescribed number. Whenever all the families of a city reach their full quota, the adults in excess of that number help to make up the deficient population of other cities.

And if the population throughout the island should happen to swell above the fixed quotas, they enroll citizens out of every city and, on the mainland nearest them . . . they found a colony . . . if ever any misfortune so diminished the number in any of their cities that it cannot be made up out of other parts of the island . . . they are filled up by citizens returning from colonial territory. . . .[11]

This idea of colonies as a solution to over-population appealed to Burton as well.[12] Obviously, then, in the utopian lands the authorities would control and manipulate the size of the population to ensure that no tenements would be erected to ruin the plans of the city-founders and to create undesirable slums.

Then comes the issue of shape. The Renaissance utopists loved symmetry. Most of the plans for ideal towns are circular, with inner and outer rings (according to the Italian tradition of the Quattrocento), except for two: Andreae who, following the German tradition that had begun with Dürer, gave Christianopolis the shape of a square; and Doni who adopted the shape of Sforzinda:[13] 'A large city, which is built in absolute perfection in the form of a star', with one hundred streets leading from one hundred gates to the temple in the centre.[14] But whatever the shape, some features of the ideal cities are constant:

(a) Streets are to be wide, straight, paved, well-lit, and so designed as to protect from the winds. In Christianopolis and Evandria special shelters are to be added along the streets for people to hide in from rain and snow.

(b) In each city there would be some public buildings, magnificent but not ostentatious.

(c) Bridges are to be built of stone and not of wood.

(d) Piazzas and fountains to please the eye and to serve practical and social functions.

11 More, pp. 75-6.
12 Burton, p. 120.
13 For the shape of ideal cities see H. Rosenau, *The Ideal City: In Its Architectural Evolution*, London, 1959.
14 Doni, pp. 6-7.

(e) Houses would be uniform, modestly but well built. They are to be made of stone, brick or flint. Windows would be glazed or covered with transparent cloth. Openings on two sides would let in plenty of fresh air and sunshine. Cellars are recommended as a measure against dampness.

(f) Gardens are to be the pride of all utopians. Fruit, vegetables and flowers are to be grown in them. In Utopia gardening competitions are to be conducted; and in Christianopolis they believe that 'the gardens are conducive to health and furnish fragrance'.

(g) Fire precautions are to be adopted in these cities: not only are all buildings to be made of non-inflammable materials, but they would also have fire-proof walls to separate the buildings (Andreae), and 'armouries in which shall be kept engines for quenching the fire . . .' (Burton).[15]

It is not too difficult to imagine how such strict plans would be adhered to in the communist utopian states where no private initiative was to be allowed and where the inhabitants would not own even the houses in which they lived. It is more interesting to note how it was considered possible to maintain such rigour in the utopias that retained private property and class differences. Agostini gave the most detailed exposition of such a plan: he believed that slums developed into hotbeds of infectious diseases because the rich lived on top of the hill, segregated from the poor. The poor, left to themselves, were incapable of living decently. Therefore in his imaginary town the rich and the poor would live in the same street, in houses built by the state. If a rich person were to desire a palace instead of a small house, he would have to submit his plan to the state's architects for approval. He would not be allowed to build higher than the rest, nor to occupy a wider space on the street. The only way to enlarge a house would be in the back courtyard up to the next street. And, of course, he would have to pay for it out of his own purse.[16]

All these lovely garden-cities would be supplied with plenty of fresh, non-polluted water. In most cases this is simply stated as a fact without an explanation as to how it would be achieved. The few utopists who tried to suggest methods were none-too imaginative. Thomas More was well acquainted with these problems in his capacities as Under-Sheriff of London and Commissioner of Sewers. In the capital of Utopia, Amaurote, water was brought from the fountain-head and 'distributed

15 More, pp. 63-6; Eberlin, pp. 126-7; Doni, pp. 6-7; Patrizi, pp. 126-8; Stiblinus, pp. 81-2; Agostini, p. 84; Campanella, pp. 217-20; *Antangil*, pp. 48-50, 93-9; Andreae, pp. 149-50; Burton, pp. 110-11, 115.
16 Agostini, pp. 84-6.

by conduits made of baked clay into the various parts of the lower town. Where the ground makes that course impossible, the rain water collected in capacious cisterns is just as useful.'[17] In the next century Burton was able to think only in terms of the existing system, but he described it as efficient and successful: 'I will have conduits of sweet and good water, aptly disposed in each town . . .'[18] Campanella added a little:

In every street of the different rings there are suitable fountains, which send forth their water by means of canals, the water being drawn up from nearly the bottom of the mountain by the sole movement of a cleverly contrived handle. There is water in fountains and cisterns, whither the rain-water collected from the roofs of the houses is brought through pipes full of sand.[19]

Zuccolo emphasized the importance of supervision and inspection:

Stewards . . . will have their eyes constantly open to see that the locks, the conduits and the canals of water are not polluted or blocked as they run through the city and the village in great numbers for the purpose of drinking, of grinding wheat, of dying cloth, of cutting stones, and for other diverse uses, and mainly for bathing.[20]

Only Andreae dared visualize a town where water is available in every house: 'An abundant supply of very clear water has been introduced into the city which they have distributed first into the streets and then into the houses, so that water abounds everywhere, and can be obtained close at hand.'[21] Overshadowed by the great Roman aqueducts, the projects of the utopists for solving the water problems of the European cities do not seem very impressive — they are but another example of the technological helplessness of the period. Yet, the insistence of them all on discussing these issues is but further evidence of how grave these problems were.

The same is true of their treatment of the question of drainage and sewage — undoubtedly the worst curse afflicting the towns. It is hardly surprising that the plague was attributed to the corruption of the air, or that aromatic fragrances were considered to be prophylactics and cures in these evil-smelling places. And yet the utopists had little to offer by way of a remedy. In the sixteenth-century only Agostini mentioned sewers in which the rain-water would run and carry all the filth out of the city and into the sea; and in times of dryness, water from the river would be run through these sewers and canals to rinse them thoroughly three or four times a month.[22]

[17] More, p. 65.
[18] Burton, p. 111; the same also in Agostini, p. 86.
[19] Campanella, in Henry Morley (ed.), *Ideal Commonwealths*, p. 234.
[20] Zuccolo, p. 69.
[21] Andreae, p. 269.
[22] Agostini, p. 86.

And in the early seventeenth century it was again only Andreae who thought up a 'modern' contraption: 'by means of underground canals they have conducted the outlet streams of a lake through the alleys, so that the flowing water frees the houses of their daily accumulations, a scheme that is more conducive to public health than anything else easily thought of'.[23]

It was easier to visualize solutions to other public 'nuisances' where only planning and legislation were involved. In Burton's utopia:

In each town these several tradesmen shall be so aptly disposed as they shall free the rest from danger or offence: fire-trades as smiths, forge-men, brewers, bakers, metal-men, etc. shall dwell apart by themselves: dyers, tanners, fell-mongers, and such as use water, in convenient places by themselves: noisome or fulsome for bad smells, as butchers' slaughter-houses, chandlers, curriers, in remote places and some back lanes.

The same principle of segregation according to trades was applied in Wolfaria and in Christianopolis, whereas in Doni's city each street would be reserved for two related trades.[24] For the same reasons the slaughterhouses in More's Utopia would be placed outside the city walls, and it would also be forbidden to bring into the city 'anything filthy or unclean for fear that the air, tainted by putrefaction, should engender disease '.[25] Without adequate means for preserving perishable foods, markets in that age were considered to be dangerous foci of disease, the warning being given by the odours of putrefaction. Both More and Agostini emphasized the prohibition on selling anything which was not absolutely fresh and wholesome and imposed the gravest penalties for doing so.[26]

Inspectors of public health are to be seen everywhere in these ideal cities. They are to inspect the state of the buildings, the cleanliness of streets and water canals, the freshness of the food in the markets. But at the same time not a word is mentioned in the utopias as to who would actually do the 'dirty work' of cleaning and disposing of rubbish. Garbage disposal presented a difficult problem to utopists of all periods. This indispensable function always had a degrading stigma attached to it. Scavengers were the lowest of the low in every society. Utopists were able to glorify all manual labour, but not scavenging. Clever devices for disposing of rubbish without the aid of human hands were beyond the imagination of utopists in previous centuries. In those utopias where the economy was not based on division of labour but on rotation, a solution

[23] Andrea, p. 269.
[24] Burton, pp. 114-15; Eberlin, p. 126; Andreae, p. 150; Doni, p. 6.
[25] More, p. 78.
[26] More, loc. cit.; Agostini, p. 87.

could be found. Such was the case in the Hellenistic utopia of Iambulus (c.250 BC), or in Bellamy's *Looking Backward* (1887) where every citizen at the age of twenty-one is recruited for three years of menial work, and only then allowed to choose an occupation according to his taste. Another possibility was offered by Fourier (1829): he believed that most boys and some girls 'have a penchant for filth . . . they love to wallow in mire', and this natureal tendency could be exploited for a practical purpose in the phalanx.[27]

The absence of scavengers in the utopias of the Renaissance is conspicuous. This institution had already been introduced in all European towns, and yet in the utopias no explanation is given as to who would sweep or rinse the streets, who would keep public grounds clean and who would dispose of the garbage. One could assume perhaps that in More's utopia the slaves not only slaughter animals, but also do all other menial jobs; or that in all utopian lands each citizen would be responsible for the cleanliness of his own home and the street in front of his house. But these are mere speculations, not corroborated explicitly in the texts. It seems that the utopists preferred not to dwell on this subject, not because they were unaware of the problem, nor — for once — because of lack of technological imagination, but because it presented an embarrassing social question which they felt was better left ignored.

Finally, quarantine. Citizens of utopian countries are usually wary of strangers. Travel abroad would rarely be allowed, not only for fear of 'cosmopolitism' (Agostini), but also for health reasons: 'The men of Evandria will not be allowed to leave . . . so that they should not bring home on their return corrupt customs, wicked habits, new maladies.'[28] When foreigners reach the shores of the utopia, they are subjected, kindly and politely, to a thorough inspection. Some would not be permitted to enter: 'Foreign vagabonds will not be allowed to enter Evandria lest they bring leprosy, mange, scabs, plague and other diseases.'[29] Others would enter but, if ill, be quarantined for a number of days. A Conservator of Health would prohibit contact of citizens with newcomers until the period of quarantine was over.[30] In this the utopists were not inventing anything original, since similar measures had been

[27] J. Beecher and R. Bienvenu (eds.), *The Utopian Vision of Charles Fourier. Selected Texts on Work, Love and Passionate attraction,* Boston, 1971, pp. 317-18.

[28] Zuccolo, p.63.

[29] Ibid., p. 48.

[30] Bacon, p. 260.

adopted by the port-towns of Europe from the days of the Black Death in the fourteenth century.[31]

(2) *Personal Hygiene and Daily Conduct*

In medieval and Renaissance culture few traces were left of the Greek and Roman cult of personal cleanliness. Soap was an expensive luxury, and its substitutes had a nauseating smell. Washing at home was a complicated ordeal and, with the spread of the syphilis epidemic, the number of public baths was greatly reduced. In any case, there was no social pressure to keep body and clothes clean — pungent perfumes were used instead. Not so in the utopias: living in a much improved environment, the utopians would be also much cleaner than the people of Europe. A sufficient number of public baths, with provision for the separation of the sexes, would exist in most of them; in Christianopolis there is even a bathroom in each home and only young children wash in public baths.[32] However, not only facilities are to be provided but also a strict routine would be dictated to all utopians:

In the morning, when they have all risen, they comb their hair and wash their faces and hands with cold water. Then they chew thyme or rock parsley or fennel, or rub their hands with these plants. . . .

They wash their bodies often, according as the doctor and master command.[33]

And although not all utopists gave such detailed accounts of the rules of hygiene, they all described with delight the fragrance of cleanness that their imaginary creatures would exude. Their clothes too are to be uniform, very modest and practical, but most of all neat and clean: 'there are wash houses for washing clothes and linen'.[34]

Cleanliness of body and garments was the obvious aspect of personal hygiene; but the utopists were concerned also with every other detail of behaviour which could affect the health of their imaginary citizens. And in this, as in other matters, they adhered to the prevalent theories of the day. A basic concept of Galenism was that, in order to maintain the normal equilibrium of the humours, a person had to follow a prescribed regimen of food, drink, sleep, exercise, and sexual activity. In the sixteenth and seventeenth centuries this idea was very popular and a great number of texts were published preaching and advocating

---

[31] 'Before the middle of the fifteenth century, the concepts of quarantine and sanitary cordons were fully developed' — Cipolla, op. cit., p. 28. On methods of quarantine see also Cipolla's earlier work, *Cristofano and the Plague*, London, 1973.

[32] Andreae, p. 152.

[33] Campanella, pp. 151, 251; Zuccolo, p. 69.

[34] Andreae, p. 153.

methods to ensure good health by adhering to a few simple rules. Among the most popular were Andrew Boorde's *Breviary of Health* (1542) and Luigi Cornaro's *Discourses* (1558). These writers and others expressed the belief that a correct diet could also prolong life. It is hardly surprising that such ideas gained a wide audience in a world where the average life expectancy was less than forty years. The belief in the possibility of prolonging life was destined to decline in later generations, but it was to remain a feature of literary utopias. In our texts we have therefore a combination of popular notions of the period together with the universal utopian desire for immortality or at least extraordinary longevity.

A major item in the utopian health programme is work. The aim was to correct the evils of both extremes: the complete idleness of the rich and the heavy strain on the working classes. 'In Naples there exist seventy thousand souls, and out of these scarcely ten or fifteen thousand do any work, and they are always lean from overwork and are getting weaker every day. They rest become a prey to idleness, avarice, ill-health, lasciviousness, usury and other vices. . . .'[35] In the utopias, however, labour would be a duty imposed on everyone (even though not always equally), and the working day would be considerably shortened.

. . . in the City of the Sun, while duty and work is distributed among all, it only falls to each one to work for about *four hours* every day.[36]

In Utopia they . . . assign only *six hours* to work. There are three before noon, after which they go to dinner. After dinner, when they have rested for two hours in the afternoon, they again give three to work and finish up with supper.[37]

For while among us one is worn out by the fatigue of an effort, with them the powers are reinforced by a perfect balance of work and leisure so that they never approach a piece of work without alacrity. . . . They have *very few working hours* . . . they all together attend to their labors in such a way that they seem to benefit rather than harm their physical bodies.[38]

None shall be over-tired but have their times of recreations and holidays;[39]

and even in Stiblinus's island they have no need to work too much since they are satisfied with so little.[40]

The utopians were to be told what to do with their free time as well. They would sleep only seven or eight hours at night on one hard mattress;[41] the rest of their time would be devoted to healthy recreations

[35] Campanella, p. 237.
[36] Ibid., p. 238.
[37] More, p. 70.
[38] Andreae, pp. 155, 161.
[39] Burton, p. 118.
[40] Stiblinus, p. 117.
[41] More, p. 60; Agostini, p. 94.

according to the tastes of the writer: prayer, study, conversation, rest, harmless games, and physical exercise.

Proper exercise, they believed, was conducive to health and beauty: 'when the women are exercised they get a clear complexion, and become strong of limb, tall and agile . . .'[42] In some utopian lands the influence of the Greek gymnasium, or of its idealization, is obvious: in the City of the Sun naked men and women exercise together, while in Evandria they play 'all the games which were practised in Rome and in Greece in order to conserve and to refine the speed, the dexterity and the strength of the body'.[43] Admittedly though, in these places, as well as in Patrizi's happy city and in Antangil, the purpose of all this vigorous exercising was not only good health but also military training: '. . . exercises which serve as good training for the profession of war and make men agile and robust in person'.[44] In More's Utopia, on the other hand, physical exercise for school children was combined with the practical purpose of helping with farm-work.[45] Burton apparently only saw the aspects of pleasure in sports, and it was only Andreae who thought of exercise solely as pertaining to health.[46]

Needless to say, no one in the utopian lands would be allowed to go hungry. On the other hand, these are not to be paradises of gluttony. The diet of the utopians would be the responsibility of the medical profession. Doctors are to prescribe the amount and the kind of food according to the season of the year and the physical condition of the person concerned. The young, the old, and the sick would get special foods. All the ingredients would always be fresh, properly cooked, and, in a very pious community, 'seasoned with wise and pious words'.[47]

Their food consists of flesh, butter, honey, cheese, garden herbs, and vegetables of various kinds. They observe the difference between useful and harmful foods, and for this they employ the science of medicine. They always change their food. First they eat flesh, then fish, then afterwards they go back to flesh, and nature is never incommoded or weakened. The old people use more digestible kinds of food, and they take three meals a day, eating only a little. But the general community eat twice and the boys four times, that they might satisfy nature.[48]

The key word in all matters of food and drink is moderation. There would be no 'gargantuan' feasts, because continence is the 'safest

---

[42] Campanella, pp. 236-40.
[43] Zuccolo, p. 66.
[44] Zuccolo, p. 49; also in Campanella, p. 241; Patrizi, p. 139; *Antangil, passim.*
[45] More, p. 68.
[46] Andreae, p. 245.
[47] Ibid., p. 159.
[48] Campanella, p. 250.

precaution for health'.[49] The utopian governments would not leave the decisions in matters of food consumption to the discretion of the individual; in different forms they would have the power of control: meals would all be taken in common (More); the food would be cooked in public kitchens (Agostini); eaten in one of the restaurants that all serve the same frugal food (Doni); or, if cooked at home, the ingredients for the meal are to be collected from a public storehouse where each family would get the exact quantities according to the number of persons sitting at her table (Andreae).

Moderation, however, prohibits not only excessive consumption, but also excessive asceticism: 'But to despise the beauty of form, to impair the strength of the body, to turn nimbleness into sluggishness, to exhaust the body by fasts, to injure one's health, and to reject all the other favors of nature . . . this attitude they think is extreme madness.'[50] Regardless of their denomination all utopists were cautious in their demand for fasts, [51] when they did, it was mainly for reasons of health; even Agostini says: 'We shall order that the entire population, twice a week, would have to be satisfied with one meal . . . which would be a way of evacuating the superfluity of the humors.'[52] Unlike many of the nineteenth-century utopists, our writers would not impose total abstinence from wine on their imaginary creatures. They would only restrict the quantities and tell the women and the children to dilute their wine with water. Obviously, the fear of alcohol as a threat to health and to society had not as yet caught on.

One thought concerning the quality of food would have delighted modern-day advocates of natural health foods: In the City of the Sun, 'They do not use dung and filth for manuring the fields, thinking that the fruit contracts something of their rottenness, and when eaten gives a short and poor subsistence, as women who are beautiful with rouge from want of exercise bring forth feeble offspring. Wherefore they do not as it were paint the earth . . .'[53]

Among those lands of frugality, Bacon's New Atlantis stands out as the one utopia of opulence and luxury. Food is a major item on his list of wonders achieved by scientific research: 'divers drinks, breads, and meats, rare and of special effects . . . sweetmeats, dry and moist, and divers pleasant wines, milk, broths and salads . . . some meats also, and

[49] Andreae, p. 245.
[50] More, p. 102.
[51] Eberlin, p. 109.
[52] Agostini, p. 98.
[53] Campanella, p. 248.

breads and drinks, which taken by men enable them to fast long after . . .'[54] Longevity in New Atlantis would not be attained through a sensible diet and a simple way of life, but by the miraculous Water of Paradise. In this Bacon combined his vision of the future achievements of science with the mythological Well of Youth or Well of Life which was a characteristic feature of Lands of Cokaygne. Such elements of 'science fiction', however, are the sole property of this one utopia.

A final aspect of personal hygiene concerns sexual activity. Only two of these utopists discussed the subject as part of the issues of health, and they represent two extreme poles. The devout Lutheran, Andreae, adopted the conventional 'puritan' attitude: 'They have the greatest desire for conjugal chastity, and they set a premium upon it, that they may not injure, or weaken themselves by too frequent intercourse',[55] while in the Platonic-pagan City of the Sun, 'some are permitted access to barren or pregnant women . . . lest they will seek some illicit outlet. Elderly mistresses and masters provide facilities for the more passionate and stimulated ones when they receive requests from them secretly or become aware of such need in the gymnasium.'[56] (Campanella's ideas on this subject were startling not only to his contemporaries but even more so to the nineteenth-century public. Thomas M. Halliday, who was the first translator of the *City of the Sun* into English in 1885, completely deleted any 'sexual' remarks. These were described by the editor as 'one or two omissions of detail which can well be spared'.)[57]

(3) *Stages in Life*

In the planned society no scope is left for spontaneity or chance. We saw how the utopists, in their eagerness to create a race of healthy people, would dictate a daily routine of work, rest, exercise, keeping body and clothes clean, regular nourishing meals, and even regulate copulation. It is therefore not surprising to discover that they would also supervise and interfere in each formative stage in the life of the utopians from infancy to death.

Patrizi and Campanella, following the Platonic tradition, went so far as to advocate eugenic breeding. Under medical supervision (with the assistance of astrologers too, in the City of the Sun), only the healthiest would be allowed to procreate; a matching of opposite types was

[54] Bacon, pp. 291-2, 295.
[55] Andreae, p. 262.
[56] Campanella, in Negley and Patrick, p. 329.
[57] Henry Morley (ed.), *Ideal Commonwealths*, p. 8.

recommended to create an ideal average; the couple would be told what to eat before and what to think and pray during the act in order to produce a perfect child. For the same purpose Campanella also recommended that women during pregnancy should observe statues of great heroes.[58]

But even the more conservative utopists, although retaining the conventional family structure, had some proposals to make to ensure the birth of healthier babies: '. . . except they be dismembered, or grievously deformed, infirm or visited with some enormous hereditary disease in body or mind; in such cases, upon great pain or mulct man or woman shall not marry'.[59] All the others stipulated the right age for marriage for both men and women. Thomas More proposed a mutual inspection of the naked bride and groom, an idea which was later adopted by Bacon as well, only he preferred that it should be done by proxy.[60]

Doni, perhaps with the Spartan myth in mind, had a different suggestion as to how to have only healthy babies:

—But what about the monsters that were born: hunchbacks, lame, one-eyed, etc., etc. Where? Where?

—There was a great well into which they threw all of them as soon as they were born, so that such deformities were never seen in that world.[61]

But on this subject, as on a few others, Doni's startlingly drastic measures were, in fact, an ironic way of expressing helplessness and frustration.

'The crowning accomplishment of women,' says Andreae, 'is bearing children, in which they take precedence of all the athletes of the earth. . . . It is certainly short of the miraculous for a woman to bear such pains, and for the child to survive the great dangers.' Therefore in Christianopolis midwives are highly qualified and held in great esteem.[62] In all other utopias too, pregnant women receive special attention, nourishing food, and good medical care. In the dining-halls of Utopia special arrangements are made so that pregnant women would be able to

---

[58] Patrizi, p. 138; Campanella, *La Città del Sole*, ed. Norberto Bobbio, pp. 61, 71; needless to say these paragraphs also were omitted from the Victorian English translation.

[59] Burton, p. 120.

[60] More, p. 110; Bacon, p. xxi. Sir Arthur Salusbury MacNalty, who described Thomas More in superlatives as the great forerunner of all public-health reformers, was wrong, I believe, in attributing to More a proposal of eugenic breeding. More's suggestions were designed mainly to prevent unhappy marriages. (Cf. MacNalty, 'Sir Thomas More as a Student of Medicine and Public Health Reformer' in E. Ashworth Underwood (ed.), *Science, Medicine and History*, London, 1953, pp. 418-36; id., *The History of State Medicine in England*, London, 1948; and in other articles. MacNalty's remarks about eugenics in Utopia were taken up, uncritically, by others as well.)

[61] Doni, p. 10.

[62] Andreae, p. 261.

retire to the cosy nurseries where 'the nurses sit separately with the infants in a dining room assigned for the purpose, never without a fire and a supply of clean water nor without cradles. . . .'[63] In all utopias babies are to be breast-fed by the mothers, and nurses recruited only if the mother dies or is taken ill.[64]

The old, the infirm, and the helpless in Renaissance Europe were dependent on the mercies of their families, of the church and of charity. And although these institutions fulfilled their voluntary functions to an amazing extent,[65] nevertheless the plight of the innumerable 'impotent poor' who could not support themselves was extremely grave.[66] In the utopian lands, all of them would become the responsibility of the state — a thought that was not destined to become reality until the emergence of the welfare states in the twentieth century. The hospitals for the utopian invalids will be discussed below, but two points should be emphasized here — the respect for old age: 'The aged of both sexes stand in the highest esteem and so they take special care that they be not afflicted with any trouble, as old age is in itself a disease. So they have appointed people who nurse them, cheer them up, honor and consult them',[67] and the suggestion that even the infirm can be of some use to society and made to feel useful: 'This thing, however, existing among them is excellent and worthy of imitation — viz, that no physical defect renders a man incapable of being serviceable except the decrepitude of old age, since even the deformed are useful for consultation.'[68]

Even death would be a more pleasant affair in the utopian lands than in reality. People would die in the hospitals. No money would be extorted from them for the last religious rites.[69] Against all accepted principles of his day, More visualized a society where euthanasia would be practised to save the incurably sick from prolonged death agonies and to rid society of such a burden. In Utopia it would be done only voluntarily and very kindly; in Doni's wise and mad world, on the other hand, it would be much more perfunctory:

— Of the incurable infirmities such as cancer, the French disease, fistula, boils, phthisis, and other maladies?

[63] More, p. 79.

[64] Loc. cit.; Patrizi, p. 139; and Campanella, p. 234.

[65] Cf. W. K. Jordan, *Philanthropy in England 1480-1660,* London, 1959.

[66] 'Philip Stubbes, writing in 1583, complained that the hospitals, lazar houses, and almshouses in some towns, cities and other places, were "not one hundred part of those that want".' (*Anatomie of Abuses*, quoted in S. R. Burstein, 'Care of the Aged in England — From Medieval Times to the End of the 16th Century', *Bulletin of the History of Medicine,* 1948, p. 743.)

[67] Andreae, pp. 270-1.

[68] Campanella, p. 239.

[69] Doni, p. 14; Eberlin, p. 114.

— Certain healing beverages, of hemlock, and arsenic and similar drugs will cure them in one hour.[70]

But dead bodies too can be a health hazard, and the utopists were quite aware of the dangers involved in the burial practices of their day. For reasons of sanitation Burton would have 'separate places to bury the dead in, not in churchyards',[71] while Campanella went so far as to imagine a society where 'they do not bury dead bodies, but burn them, so that the plague may not arise from them'.[72] Even in Utopia, it seems, cremation would sometimes by practised along with conventional burial and the unceremonious drowning of the bodies of suicides.[73]

B. REMEDIAL MEDICINE

Despite all the preventive safeguards, the utopists did not suppose that disease would disappear altogether, and therefore set out to describe a detailed progamme for medical services that they believed would be much better than the existing institutions and facilities.

Zuccolo was the sole utopist to decide that the perfect way of life in his imaginary world would make doctors redundant: 'Neither physicians nor apothecaries would be tolerated in Evandria.'[74] Patrizi, on the other hand, admitted their necessity: our health, he says, may suffer from many various causes which cannot be avoided, and therefore it would be necessary to maintain in the Happy City another group of artisans ('artefici') — physicians, surgeons, barbers, and apothecaries. But he seems to include them with reluctance and relegates them to the lowest ranks of his rigid hierarchy.[75] Zuccolo and Patrizi, however, are the exception; the rest of our writers attributed to the medical profession a highly important role in the perfect society. In comparison with the wrath poured on the legal profession,[76] the doctors of this world received a relatively lenient treatment from the utopists. Their greed was implicitly criticized, as was their fierce struggle to reserve for themselves a monopoly in the field of healing; but little was said about the greatest

[70] More, pp. 108-9; Doni, p. 10. As mentioned above (p. 21), Doni was influenced by Guevara. In The Diall of Princes (English translation, London, 1557, fol. 47) the seventh law of the Garamanti is as follows: 'We ordaine that no woman live above 40 years, and that the man lyve until 50, if they dye not before that time, and then they be sacrificed to the Gods: for it is a great occasion for men to be vycious, to thinke that they shall lyve many yeres.'

[71] Burton, p. 111.
[72] Campanella, p. 260.
[73] More, p. 109.
[74] Zuccolo, p. 57.
[75] Patrizi, p. 128.
[76] See ch. 5.

of their faults — ignorance. This, the utopists realized, could not be easily corrected. Nevertheless from the utopias emerges a programme for far-reaching reforms in the *organization* of the medical services, which is one of the most striking features of these ideal societies.

First and foremost, in all utopian lands doctors are to be salaried employees of the state and the entire population — both in the communist and the non-communist utopias — would receive free and equal medical care: 'chirurgeons and physicians, which are allowed, to be maintained out of the common treasure, no fees to be given or taken upon pain of losing their places, or if they do, very small fees'.[77] 'Famous physicians shall be sent for and paid from the common purse.'[78]

Admittedly, medical care in those days could be an additional hazard to life rather than a benefit. Furthermore, in some parts of Europe the ratio of university-trained physicians to the population was relatively high and, therefore, their services were not limited — as is often supposed — to the aristocracy and the rich. Some towns also provided public funds to hire medical men to serve as community doctors and to take care of the poor (e.g. the *medici del pubblico* in Italy, where such policies were practised more vigorously than anywhere else).[79] Nevertheless, free medical care for all was obviously an extremely radical aspiration.

In the City of the Sun doctors are practically the rulers of the town. Under the supervision of Mor they control most aspects of daily life: hygiene, food, clothes, public sanitation, and — most of all — procreation. They and the astrologers influence the life of the people more than anybody else.

Much thought and space is devoted to the medical profession, its organization and training, in the *Repubblica Immaginaria*: in every quarter of this city there would be a medical team consisting of a physician, a surgeon, and an apothecary. There would be also two shops to provide 'tutte le cose che siano necessarie al vitto umano'. An apprenticeship of five years would be required of the doctors, not only for the purpose of learning the practice of medicine, but also in order to make the young physicians familiar with the 'complessiono degli uomini di quel quartiero'. The medical team would go also to visit the sick in the villages outside the city. But a citizen would have the right to consult other doctors if he was not pleased with those assigned to his quarter.[80]

[77] Burton, p. 116.
[78] Eberlin, p. 129.
[79] See Cipolla, *Public Health*, pp. 83, 87; and *Cristofano*, p. 45.
[80] Agostini, pp. 99-100.

The most 'modern' health service is to be found, oddly enough, in the 'reactionary' utopia, *Antangil*. One point was emphasized by the anonymous author very strongly: the need to abolish the harmful separation between the branches of the medical profession. The antagonism and friction between physicians, surgeons, barbers, and apothecaries were apparently strongest in sixteenth-century France, although existing elsewhere in Europe as well.[81] The author resented this state of affairs not only because of the wasted energies spent on retaining monopolies, but mainly because he thought it outrageous that qualified physicians had hardly any practical knowledge of the human body. 'Those doctors that have only some theory and are without perfect knowledge of the causes and proper remedies of diseases, commit great errors which are injurious to the infirm.'[82] In Antangil, therefore, a doctor would be a surgeon and an apothecary as well as a physician. Using a scalpel, performing operations, and mixing drugs would not be considered menial tasks. Medical students would be taught the practical with the theoretical aspects of the science, and they too would have to undergo an apprenticeship of five years. Only at the age of thirty would they be allowed to start an independent practice. Even in this stratified society, with its highly privileged aristocracy, doctors would receive salaries from the state, and be compelled to treat the poor gratis, and from the rich to demand only a small fixed sum. The result would be an excellent, well-qualified profession, and the aristocracy would no longer find any use for foreign charlatans.[83] And as doctors become state employees, hospitals become public institutions maintained and administered by the government.

Hospitals of all kinds, for children, orphans, old folk, sick men, mad men, soldiers, pest-houses, etc., not built *precario* by gouty benefactors . . . not by collections, benevolences, donaries for a set number (as in ours), just so many and no more at such a rate, but for all those who stand in need, be they more or less, and that *ex publico aerario*, and so still maintained.[84]

Burton, Eberlin, and Agostini use 'hospital' in its traditional wide sense to cover every type of 'home' and almshouse:

---

[81] See Alfred Franklin, *La Vie privée d'autrefois*, Paris, 1890, Vol. 1 — *Les Médecins*, Vol. 2 — *Les Chirurgiens*. This work is a fascinating description of everyday life in these centuries. Two other volumes in the series, which are relevant to this chapter are: Vol. 7 — *Les Médicaments*, and Vol. 10 — *L'Hygiène*.

[82] *Antangil*, pp. 138-9.

[83] Ibid., pp. 138-40.

[84] Burton, p. 111.

If there are more than two monasteries and two convents in one place, they shall be dissolved and turned into hospitals for the poor old men who served a town as officials, or for their sick children. Such hospitals shall be well designed and built. Some shall be for travellers. Some for common poor inhabitants.[85]

Indeed, in real life hospitals were associated with the poor and helpless. But in the utopias based on complete equality, where there would be no poor, hospitals would either disappear or acquire the one function they have in modern times — treatment of the sick. In the *City of the Sun* no hospitals are mentioned; in *Christianopolis* they seem to have been added as an afterthought:

Medicine. surgery, and the kitchen are all equally at the disposal of the sick, and everyone is prompt to assist. . . . Married women and widows here have the greatest opportunity and skill, and the state very kindly commends to them the care of the sick; they even have hospitals intended to take care of them.[86]

But Doni made their purpose clear and dissociated them from any connection with poverty:

— But he who fell sick?
— Went to the street of Hospitals where he was cured and visited by doctors . . .
— Oh, how dreadful that a rich man should have to go to a hospital!
— It makes sense; there no one was richer than another.[87]

In Antangil there would be a hospital for the local sick in each town and village, suited to its size, and a doctor with two assistants would work in each of them. But the longest and the most modern description of hospitals appears in the first of the utopias:

They have four at the city limits, a little outside the walls. These are so roomy as to be comparable to as many small towns. The purpose is twofold: first, that the sick, however numerous, should not be packed too close together in consequent discomfort and, second, that those who would have a contagious disease likely to pass from one to another may be isolated as much as possible from the rest. These hospitals are very well furnished and equipped with everything conducive to health. Besides, such tender and careful treatment and such constant attendance of expert physicians are provided that, though no one is sent to them against his will, there is hardly anybody in the whole city who, when suffering from illness, does not prefer to be nursed there rather than at home.

The sick, as I said, are very lovingly cared for, nothing being omitted which may restore them to health, whether in the way of medicine or diet. They console the incurably diseased by sitting and conversing with them and by applying all possible alleviations.[88]

More did not forget to mention the kind treatment that would be shown to people who are mentally deficient.[89] The same treatment they

[85] Eberlin, p. 129; Agostini, p. 90.
[86] Andreae, pp. 273-4.
[87] Doni, pp. 8-9.
[88] More, pp. 78, 108.
[89] Ibid., p. 113.

receive in Christianopolis too, where there would be also institutions for violent mental patients.[90]

Thus far the organization of the medical system in utopias; in their thoughts on medical theory, however, the utopists revealed less originality. Although none of them gave a detailed exposition of doctrine, it is quite apparent that they did not go much beyond the conventions of their day. The European travellers who discovered Utopia brought with them 'some small treatises of Hippocrates and the *Ars Medica* of Galen, to which books they (the Utopians) attribute great value'.[91] The same names of the revered Fathers of Medicine appear in *Antangil* and in the utopia of Stiblinus who included also Avicenna and Nicandro.[92] Agostini would not permit innovations in the practice of medicine; his doctors would be forbidden to use 'dangerous remedies' and required 'to cure canonically'.[93] In Evandria the methods of cure, without the aid of physicians, would be the simple and the ancient: 'with diet, with bloodletting and with simple medicaments they cure even the gravest of maladies'.[94] The long list of cures given by Campanella, although quite impressive, is entirely within the tradition of humoral pathology, herbal lotions, diet, and balneotherapy. 'They heal fevers with pleasant baths and with milk-food, and with pleasant habitation in the country and by gradual exercise. . . . Fevers occurring every fourth day are cured by suddenly startling the unprepared patients, and by means of herbs producing effects opposite to the humours of this fever.'[95] And on the city walls, where all human knowledge is depicted, including liquids which cure all diseases, no space was left for the discoveries of the future.[96] In Antangil knowledge is accumulated not through medical research but by a very naïve method of collecting all 'secrets' from foreign physicans and making them public in a catalogue which is published annually.[97]

As we saw, the only criticism expressed in these texts of the education of doctors concerned their lack of practical knowledge in anatomy. This would be rectified in some of the utopian lands:

[90] Andreae, p. 274.

[91] More, p. 105.

[92] *Antangil*, p. 138; Stiblinus, p. 97; ('Nicandro' is probably Nicander of Colophon (third or second century BC). His *Theriaca* and *Alexipharmaca* are accounts of poisons and their antidotes).

[93] Agostini, p. 101.

[94] Zuccolo, p. 57.

[95] Campanella, pp. 251-3.

[96] Ibid., p. 222; Campanella, a prolific writer, devoted many pages in several of his works to the science of medicine and to the different cures; but basically on the same lines as in the *City of the Sun*. See for example his *Realis Philosophiae epilogisticae partes quatuor* (1623), in Carmelo Ottaviano (ed.), Rome, 1939, pp. 280-92, 529-45.

[97] *Antangil*, p. 139.

They have also a place given over to anatomy, that is, the dissecting of animals. . . . The inhabitants of Christianoplis teach their youth the operations of life and the various organs from the parts of the physical body. They show them the wonderful structure of the bones, for which purpose they have not a few skeletons . . . they also show the anatomy of the human body, but more rarely.[98]

Dissections of human bodies were not very common as yet, even in the utopias. Francis Bacon, for instance, only mentioned the dissections of beasts and birds that were performed in Salomon's House.[99] Campanella, however, showed less hesitation: in the City of the Sun, as in some university towns in Europe, human anatomy would be studied on the bodies of criminals condemned to death.[100]

Nevertheless some non-conservative attitudes towards medical research found expression at least in two of the utopias. Andreae did not mention the names of Hippocrates and Galen even once; influenced by the Paracelsian school, he put the emphasis on iatrochemistry and alchemy, and desired to expose the potentialities of science as a means to attain solutions to disease. In the laboratories of Christianopolis 'the properties of metals, minerals, and vegetables, and even the life of animals are examined, purified, increased and united, for the use of the human race in the interests of health'.[101] In the pharmacy of the same town one could find 'whatever the elements offer, whatever art improves, whatever all creatures furnish'.[102] Bacon too, although he admired Hippocrates and detested Paracelsus, was influenced by the chemical doctors who were not Paracelsians, and the list of achievements of Salomon's House is, in fact, an ardent attempt on Bacon's part to supply the missing link between the advancements in science made during the sixteenth century and the backward art of healing. The gardens, laboratories, and mines supply every conceivable cure: a 'scarlet orange' as an effective prophylactic against seasickness;[103] grey pills which hasten recovery; qualified air and special baths, fruits and plants, exquisite separations and distillations, as well as the miraculous Water of Paradise.[104]

---

[98] Andreae, p. 199.

[99] Bacon, p. 290.

[100] Campanella, ap. Bobbio (ed.); p. 107; Phyllis Allen, 'Medical Education in 17th Century England', *Journal of the History of Medicine*, 1946, p. 126.

[101] Andreae, p. 196.

[102] Ibid., p. 198.

[103] 'A good Sivill orenge stuck with cloves' was considered in England to be a prophylactic against the plague; again the principle is that of pleasant smells to prevent disease. (Wilson, op. cit., p. 8.)

[104] Bacon, p. 263. See discussion in Charles Webster, *The Great Instauration*, pp. 248-9.

The significant fact, however, is that the utopists, whether of the conservative school or adherents of new-fangled ideas in medicine, whether imaginative or dull, all attached great importance to the science of medicine, because they saw its direct relevance to the alleviation of human suffering. In some of them it is the one and only science studied and practised: 'because in the Republic it is considered a necessary science. Since what better refuge from the distress of disease than the advice and aid of doctors?'[105] In the others it is regarded as the most important of all disciplines:

Even though there is scarcely a nation in the whole world that needs medicine less, yet nowhere is it held in greater honor — and this for the reason that they regard the knowledge of it as one of the finest and most useful branches of philosophy.[106]

We will respect medicine, not so much because it offers us an unusually long life or sets itself against death, but because our excellent Creator has wished that through His creatures and their use, benefit should be brought to us.[107]

What would be the results of all these reforms, research, planning and regimenting? Disease would not be totally eradicated; they could not even visualize a world free of the plague, since they did not know its causes. In Christianopolis 'when a plague rages it is wonderful to relate how little effort there is to escape; they await the hand of God';[108] in Utopia whole towns are sometimes depopulated by this horrible scourge;[109] and in the imaginary city of Agostini all doctors would gather for *consilium* in hours of such emergency, unlike the doctors in real life who, in many cases, were the first to flee on such occasions.[110] Bacon could imagine that through scientific measures such disasters would be predicted before they occur, but not that they would be entirely prevented. The Solarians had not managed to overcome epilepsy — an affliction with divine connotations.[111]

Nonetheless, if not promising a paradise of absolute health, the utopists were trying to instil hope:

---

[105] Stiblinus, p. 97; it is also the sole science mentioned in *Wolfaria, Evandria*, and in Agostini's imaginary city.

[106] More, pp. 105-6.

[107] Andreae, p. 246.

[108] Ibid., p. 274.

[109] More, p. 77.

[110] Agostini, p. 100. On 12 October 1630 the Health Board of Florence addressed the following circular to officials in all towns under its jurisdiction: 'You will at once notify each of the aforesaid physicians and surgeons in writing that they must not leave the states of His Most Serene Highness on pain of their lives and confiscation of their goods.' This is but one example of the battle conducted by authorities to prevent the flight of doctors from areas contaminated by plague. (Quoted in Cipolla, *Public Health*, p. 67).

[111] Bacon, p. 298; Campanella, p. 252.

Nowhere are men's bodies more vigorous and subject to fewer diseases;
. . . there is never gout in the hands or feet, nor catarrh, nor sciatica, nor grievous colics, nor flatulency, nor hard breathing. . . . The length of their lives is generally one hundred years, but often they reach two hundred.[112]

Most of the thoughts on the issues discussed above were not original contributions of the utopists. They borrowed ideas from popular literature on diet and hygiene, non-implemented statutes of local authorities concerning sanitation, prevalent medical theories and superstitions, and what they believed to be the practices in other parts of the world. Since the Black Death of the fourteenth century and the epidemics that followed, there was a growing awareness throughout Europe of the problems of public health; the best example was the expansion of the responsibilities of the Health Boards in Italy, with their *Provveditori alla Sanità* and *medici del pubblico*, beyond mere establishment of quarantines and sanitary cordons. The utopists reflected these developments in the administration of public health, as they also expressed the helplessness of the age to find technological solutions to health hazards.

However, the important contribution of the utopists is not limited to the reflection of contemporary thoughts and problems. Each of them made his eclectic collection into a coherent system, a comprehensive plan to create a healthy society. Moreover, they put the onus entirely on the state. In all these lands it is neither the Church, nor the local authorities, nor charities, but only the central government which would be responsible for the physical well-being of the entire population. As far as medicine is concerned, the utopias of the Renaissance are all true welfare states. But together with the responsibilities, the utopists endowed the government with incredible power. Advice in matters of personal diet and hygiene would become law; ordinances would be carried out to the letter; the rulers would dictate every action and interfere in every stage in the life of their citizens in order to protect the health of the community.

The utopists refrained as much as they could from decorating their creations with features which would have been considered incredible. A world free of the plague was beyond their imagination, taps of running water in every home were an unattainable dream. But a government that interferes in the daily life of the citizens only for altruistic reasons, and a population that willingly obeys all laws — this they did not consider implausible.

---

[112] More, p. 105; Campanella, pp. 250-2.

3

THE IDEAL EDUCATION

Education as a responsibility of the state is an inherent idea in any conception of utopia. In the planned society the immensely influential power of education cannot be left to the whims of individuals — it must be organized by the central authority to ensure uniformity and correct behaviour. The utopist designs the mould in which the children of each successive generation should be cast, and he relies on the imaginary government to see to it that not a single soul escapes the mould. The state that had achieved perfection must remain static, and it cannot afford the variety and the unpredictability of education by private enterprise.

This basic utopian notion is as old as Plato's *Republic*. In the utopias of the Renaissance it is most succinctly expressed by Campanella:

For they say that children are bred for the preservation of the species and not for individual pleasure, as St. Thomas also asserts. Therefore the breeding of children has reference to the commonwealth and not to individuals, except in so far as they are constituents of the commonwealth. And since individuals for the most part bring forth children wrongly and educate them wrongly, they consider that they remove destruction from the state, and therefore, for this reason, with most sacred fear, they commit the education of the children, who as it were are the element of the republic, to the care of the magistrates.[1]

The magistrates in utopian lands take the burden of education upon themselves not for the benefit of individuals, not in order to provide 'equal opportunities', nor for any other such modern slogan, but solely in order to safeguard the stability of the state. In other words, the utopian educational system serves to produce the perfect cogs for the perfect social machine; it is hardly to be expected that it should serve to satisfy the aspirations of the individual cog.

However, in putting the onus of education on the central authority, the utopists of the Renaissance were not just following the inevitable pattern of the genre: they were also expressing criticism of the existing educational reality. Before the second half of the seventeenth century there was nowhere in Europe a system of state-organized education. The battle between Protestants and Catholics over the souls of the Europeans

[1] Campanella, pp. 235-6.

did indeed breed a deeper awareness of the importance of education on a large scale, with the resulting proliferation of theories and experimental schools. Where heads of states took over many functions of the church, they too came to realize their responsibilities in education. Yet the transition was slow, and throughout our period the educational opportunities remained quite similar to what they had been before the Reformation: a heterogeneous amalgam of court-schools and parish-schools, private tutors and charity-endowed institutions, with traditionally Scholastic or humanistically inspired curricula. In any case they left the majority of the population, if not totally illiterate, at least too ignorant — in the eyes of our utopists — to be profitably harnessed to the cause of social reform.

The demand for the provision of education by the state does not necessarily imply that it should be universal and equal (not like the demand for the provision of health care to all by the state — for it is essential to the perfect society that all its members be healthy, but it is not self-evident that they must be all learned). In three of our utopias — The Happy City, Antangil, and Evandria — learning is reserved as a privilege to the upper classes only. Education is these cases serves the purpose of perpetuating the hierarchical structure of society.

The children who are destined to learn, in all the utopias, are removed from home and educated by professional pedagogues. In two lands — Doni's and Campanella's — babes are taken from the arms of their mothers immediately when they are weaned to be deposited in the hands of special instructors. In another group of imaginary countries children are taken from the family into boarding schools at the age of six (Antangil, Christianopolis) or at the age of ten (Evandria). In the rest of the utopian states the separation of children from their parents is effected only through day-schools. The structure of the family unit would be affected accordingly. Doni and Campanella explicitly desired its abolition since they believed that family ties conflict with absolute loyalty to the state and comprise a superfluous emotional and economic burden. The others, however, had no wish to change this basic social institution. They sought to alleviate the anxiety of the parents caused by the separation from their offspring: 'No parent gives closer or more careful attention to his children than is given here, for the most upright preceptors, men as well as women, are placed over them. Moreover, they (the parents) can visit their children, even unseen by them, as often as they have leisure.'[2] Disciplinary functions were often reserved to the parents: 'Husbands

---

[2] Andreae, p. 208.

correct the wives, and parents their children . . ."[3] The extended family, with all its social significance, is elaborately glorified in the 'Feast of the Family' in New Atlantis.[4] But on the whole, the utopists did not attach to the family framework much importance in the process of education (except, as we shall see, in vocational training). In this they differed from such contemporary theorists as Vittorino da Feltre and Vives who thought that the parents could fulfil an important role in the elementary stages of the child's tutelage.[5] In the imaginary perfect societies all education would be institutionalized.

Public education is an immensely expensive enterprise. In the communist utopias it would be paid for, naturally, out of the common purse. In Wolfaria, it seems, the fortunes of the dispossessed Church were designated for this purpose, at least in the initial stages of reform. In Antangil tuition fees were to be determined according to the economic circumstances of the family, and thus the richer parents would subsidize the poorer. On the whole, however, the utopists remain vague on this practical issue, either because they were not aware of the vast proportions of their scheme, or because they did not have solutions to offer.

While agreeing on who should be responsible for providing education, out utopists differ about its purposes and aims. Although nowhere stated explicitly, the aims can be clarified by viewing the end-products of the educational process, i.e. the ideal types presented by the author as the members of his imaginary society. Three main types emerge from these texts, which I propose to name: 1. The Useful Citizen; 2. The Enlightened Ruler; 3. The Scientist. Some utopias describe the programme for producing only one of these models, others for all three. We shall examine them separately, bearing in mind that some overlapping does exist.

(1) *The Useful Citizen*

Each utopian is given four kinds of training: moral, academic, vocational, and military.

They take the greatest pains from the very first to instill into children's minds while they are still tender and pliable, good opinions which are also useful for the preservation of their commonwealth. When once they are firmly implanted in children they accompany them all through their adult lives and are of great help in watching over the condition of

---

[3] More, p. 112; also Agostini, p. 51. Zuccolo, p. 54.
[4] Bacon, pp. 279-83.
[5] See W. H. Woodward, *Vittorino da Feltre and other Humanist Educators,* Cambridge, 1897, pp. 192-5.

the commonwealth. The latter never decays except through vices which arise from wrong attitudes.[6]

This overt defence of ideological indoctrination meant, in the sixteenth century, little beyond the instillation of religious principles.[7] Thomas More describes the Utopians as being openly pragmatic in the adherence to certain religious beliefs despite the polytheistic freedom. King Utopus made the whole matter of religion an open question and left each one free to choose what he should believe. By way of exception, he conscientiously and strictly gave injunction that no one should fall so far below the dignity of human nature as to believe that souls likewise perish with the body or that the world is the mere sport of chance and not governed by any divine providence. After this life, accordingly, vices are ordained to be punished and virtue rewarded. Such is their belief, and if anyone thinks otherwise, they do not regard him even as a member of mankind. . . . Who can doubt that he will strive either to evade by craft the public laws of his country or to break them by violence in order to serve his own private desires when he has nothing to fear but laws and no hope beyond the body.[8]

Similar sentiments — that there is no social morality without religion — were expressed by Patrizi[9] and Campanella[10] as well. The rest of the utopists were less blatantly utilitarian in their attitude to the religious creed. For them a good Christian would be, *ipso facto*, a good citizen. Thus moral training in most of these lands is strictly theological. In Wolfaria schoolchildren have two lessons a day in the Evangelical commandments and on Sundays they are admonished by the chaplain. Teachers in this country would be obliged to go out into every hamlet 'to tell the peasants something edifying about God'.[11] The next devout Lutheran utopia is Christianopolis where 'Three prayers are offered each day, morning, noon, and in the evening when thanks are given to God for blessings received. . . . No one may be absent from these prayers . . . parents bring all their children hither that they may learn even in infants' prattle to praise God.'[12] A separation between the civil and the ecclesiastical authorities was maintained only in Agostini's utopia. In this post-Tridentine Catholic land all education was to be provided by the church. From the little that Agostini says on this matter it would seem that religious–moral training forms the only content of schooling. When Zuccolo describes the instruction of the youth in Evandria by the Pendonomi, he too only mentions that 'they train them to fear and revere God, to obey the magistrates, to honour their fathers and

[6] More, p. 140.
[7] See, for example G. Strauss, *Luther's House of Learning: Indoctrination of the Young in the German Reformation,* Johns Hopkins University Press, 1980.
[8] More, pp. 134-5.
[9] Patrizi, p. 134.
[10] Campanella, p. 261.
[11] Eberlin, p. 108.
[12] Andreae, p. 158.

mothers, to respect old and learned men, to love and cherish friends, companions and relatives, not to lie, not to cheat, not to offend'.[13] Perhaps because of the congruence of ideology and religion in this period, the amount of non-religious ideological instruction in the utopias is practically nil. To the modern reader, who has learnt to expect rigorous indoctrination in every ideologically inspired society, it comes as a surprise to discover that moral training in these ideal communities is not different in any significant way from the usual precepts taught to children or to adolescents in every 'open' society. But — as we shall try to indicate below — the inculcation of 'correct' moral and ideological ideas in the utopian population would be done more by prohibitive measures rather than by positive indoctrination.

Most of the utopists agree that the ordinary citizen should receive at least some elementary academic education. With the exception of Patrizi's Happy City, no utopian country would have illiterate citizens. Needless to say, an entirely literate society, no less than free and equal health services, was an extremely radical aspiration in the sixteenth and seventeenth centuries.

All children are introduced to good literature.[14]

All children, boys and girls shall go to school when they are three years old. They shall stay there until they are eight.[15]

Both sexes are instructed in all the arts together.[16]

All the children of citizens in general, children of both sexes, are taken into training.[17]

The school ages might vary: from three to eight, from six to fourteen, from ten to eighteen, but basically they were all advocating compulsory universal primary education.

However, when it comes to the curriculum of this elementary schooling, the utopists had little innovation to offer. The subjects of study, though they vary slightly from utopia to utopia, are not much different from the accepted syllabus in many of the existing schools. The minimal programme is offered in the kingdom of Antangil (for those who would not attend the Academy): 'In each parish there would be a master . . . they would be taught first of all to read well, to write, draw and calculate, with a few moral precepts, together with the catechism and principal points of the Christian faith.'[18] In Wolfaria the list is slightly longer:

13 Zuccolo, p. 48.
14 More, p. 89.
15 Eberlin, p. 127.
16 Campanella, p. 227.
17 Andreae, p. 208.
18 *Antangil*, p. 140.

At school children will learn to understand Latin as well as German, they shall have a superficial knowledge of reading and writing in Greek and Hebrew. . . . All children will learn the arts of measuring, counting and astronomy. All children will learn to distinguish common herbs and common medicines, against common illnesses.[19]

Stiblinus, himself a teacher, who dreamed of a society in which all citizens were men of letters, proposed the fullest Humanistic programme of studies for all the youth. They will be taught non-scholastic philosophy, Greek literature, mathematics, and music, in addition to all the basic subjects.

Actually, our utopists found themselves facing a dilemma when it came to the extent and the content of learning for the entire population. They felt, on the one hand, that learning was a luxury, that the 'useful cogs' needed very little academic training in order to fulfil their role in society; yet, on the other hand, the humanists amongst them, could not conceive as ideal a society whose members were not all ardent seekers of culture. Hence the idealized creatures of More's imagination who spend all their free time in 'cultivating the mind'; hence the defiant statement of Andreae's:

This feature, moreover, is entirely peculiar to them, namely that their artisans are almost entirely educated men. For that which other people think is the proper characteristic of a few . . . the inhabitants argue should be attained by all individuals. They say neither the subtleness of letters is such, nor yet the difficulty of work, that one man, if given enough, cannot master both.[20]

However, the essential part of the education of the Useful Citizen is his vocational training. Only a tiny minority would be allowed to continue in academic studies as full-time scholars: the rest would be obliged to choose a profession, to master it and to excel in it. A community of highly-skilled artisans is what most of these utopists desired. But, once again, in the actual programme for vocational training they had little to offer beyond the improvement of existing methods. The common proposal was that the child should learn the trade of his parent:

For the most part, each is brought up in his father's craft, for which most have a natural inclination. But if anyone is attracted to another occupation, he is transferred by adoption to a family pursuing that craft for which he has a liking.[21]

Parents shall teach their children, one of three at least, bring up and instruct them in the mysteries of their own trade.[22]

---

[19] Eberlin, pp. 127, 130.
[20] Andreae, p. 157.
[21] More, p. 69.
[22] Burton, p. 114.

The trades in Evandria are passed, as if hereditary, from father to son; since each artisan has two or more sons, he is required by law to instruct at least one of them in his trade. . . . Thus the crafts attain an exquisite fineness.[23]

Agostini in his imaginary state would retain the guild system: all children from the age of fourteen would choose a profession and would be trained in it within the framework of one or the other of the 'universitàs'. A more advanced system appears for the first time in the utopias of the early seventeenth century. The children of the Solarians are taken at an early age to 'the offices of the trades, such as shoemaking, cooking, metal-working, carpentry, painting, etc. In order to find out the best of the genius of each one'.[24] Then they are trained in the profession of their choice by masters of that trade: 'they set that one apart for teaching the art in which he is most skilful'.[25] In Christianopolis vocational training is part of the school curriculum — also taking into account the natural inclinations of the child.[26] Later they too would be trained by those artisans who excelled in their trades.[27] This combination of a school course and then specialized training by the best in the profession is certainly the most progressive of the proposals in this particular field.

Although describing the life in the cities only, the utopian countries are basically agrarian societies, totally dependent for their livelihood on the land and on the men who work it. Two of our utopists — More and Campanella — desired to blur any sharp distinctions between the town and the country.[28] Thus in Utopia all citizens work the land in rotation for periods of two years, while in the City of the Sun all citizens are mobilized, to the sound of drums and trumpets, for agricultural work during the harvest seasons. In both these countries, therefore, children receive training in the theory and practice of agriculture as part of their schooling. In Agostini's Imaginary Republic, on the other hand, special instructors go out to the village to improve methods and to raise the level of farming.

The last item on the educational programme for the Useful Citizen is military training. The problematical issue of the existence of war in descriptions of ideal societies and, moreover, its aggressive and Machiavellian nature, had been discussed at length in almost every book

[23] Zuccolo, p. 53.
[24] Campanella, p. 227.
[25] Ibid., p. 246.
[26] Andreae, p. 210.
[27] Andreae, p. 157.
[28] An intention highly praised by the Marxist admirers of these two 'precursors of Socialism'.

or essay on utopias.[29] But whatever our value judgements in this matter, the fact remains that practically all our utopists believed in the necessity of preparing for war. In the same manner that they could not envisage a world without the plague, so were they incapable of imagining a world without war. Their utopias are small societies of perfection surrounded by an unreformed world, therefore they must be ready to protect themselves from all potential enemies. There are no professional soldiers in most of these utopian societies,[30] so that the entire population is trained to take up arms should the need arise.[31] Intensive athletic activity in these communities keeps the people healthy and fit for battle. Then they are taught the use of weapons and all other aspects of the 'science of war':

Power is at the head of all the professors of gymnastics who teach military exercise, and who are prudent generals, advanced in age. By these the boys are trained after their twelfth year. . . .

The women are also taught these arts under their own magistrates and mistresses, so that they may be able if need be to render assistance to the males in battles near the city. . . . In this respect they praise the Spartans and Amazons. . . .

Every day there is practice of arms. . . . Nor are they ever without lectures on the science of war. . . .[32]

The description of the military exercises and the theoretical discussions of historical battles in Campanella is especially long, but in shorter versions one finds accounts of the military drill in almost all the utopias. A curious exception in this orchestra of sabre-rattling is Doni. He claimed that the extreme frugality of his imaginary society would prevent jealousy among its neighbours and thus ensure its peaceful existence.[33] In his land, therefore, no one even rides a horse. Apart from Doni, the rest of the utopists put their trust in the patriotic feelings of

[29] An excellent analysis of this question, together with a comprehensive summary of historical attitudes to the chapter on warfare in More's *Utopia* is to be found in S. Avineri, 'War and Slavery in More's Utopia', *International Review of Social History*, 7 (1962), pp. 260-90.

[30] With the exception of Patrizi's state where the 'guardians' form the army, and Burton's utopia which would have a standing army. On the polemics regarding this subject see L. G. Schwoerer, *'No Standing Armies!' The Anti-army Ideology in 17th Century England*, Johns Hopkins University Press, 1975; See also Machiavelli's famous passage on the advantages of a civil militia in *The Prince* chs. XII, XIII; and the analysis of this issue in Renaissance political thought in H. Baron, *The Crisis of the Early Italian Renaissance*, Princeton University Press, 1966, pp. 430-40; and in Q. Skinner, *The Foundations of Modern Political Thought*, Cambridge University Press, 1978, Vol. I, pp. 130-1.

[31] In Antangil only the students of the Academy are trained in arms.

[32] Campanella, pp. 239-40.

[33] Campanella claimed the exact opposite: 'there are four Kingdoms in the island, which are very envious of their prosperity . . .', p. 241.

the people, their ideological conviction of superiority, universal conscription, and thorough military training; or in the words of Thomas More:

The absence of anxiety about livelihood at home, as well as the removal of that worry which troubles men about the future of their families (for such solicitude everywhere breaks the highest courage), makes their spirit so exalted and disdainful of defeat.

Moreover, their expert training in military discipline gives them confidence. Finally, their good and sound opinions, in which they have been trained from childhood both by teaching and by the good institutions of their country, give them additional courage.[34]

It needs to be emphasized that in some of the utopias the entire programme for producing useful citizens applies to both men and women. Many of these writers, particularly the advocates of communist societies, did not assume any inferiority of the female sex in intellectual capacities. Even if they were to be assigned lighter work, they would receive the same academic, vocational, and even military training as the men. 'I do not know why this sex, which is naturally no less teachable, is elsewhere excluded from literature.'[35] In the teachings of several educationalists of the Renaissance, in particular Vittorino da Feltre and Vives, the importance of learning for women was often stressed. But they had in mind only women of leisure and position, and the curriculum was aimed at giving them household training, forming their character, and enabling them to converse intelligently. The utopists, however, were concerned with all women and their training for an active role in society.[36] This, undoubtedly, is one of the most striking features of the Renaissance utopias.

(2) *The Enlightened Ruler*

The creation and the preservation of an ideal society, according to the Renaissance utopists, are totally dependent on the wisdom of the rulers. These imaginary societies were not described as evolving gradually, nor as a result of a popular revolution, but as a creation, *ex nihilo*, by an infinitely wise king or a group of highly intelligent people. Only a succession of wise and learned rulers would ensure the continuity of the social perfection. The implied assumption is, of course, that wisdom equals virtue and knowledge prevents moral defects. This absolute reliance on the Enlightened Ruler — so foreign to modern political thought — was a major characteristic of much of the contemporary

[34] More, p. 126.
[35] Andreae, p. 210.
[36] See also Ch. 4.

literature. Consequently, the question of how to create an educated ruling élite preoccupied many thinkers of the day. The utopists combined the fashion with the platonic model of the philosopher-king and thus devoted a great deal of thought to the process of educating the 'governors' of the ideal society. This for them was not only an educational problem but also political policy, since a well-educated benevolent ruler is their solution to a plethora of questions concerning the protection of the people from tyranny.

In the utopias of Patrizi, I.D.M., and Zuccolo all learning (higher than the most elementary) is reserved to the upper classes only. The élite in these lands is determined by class origin, and education is its main privilege. Patrizi explicitly declares that the working classes are too base and crude, and too fatigued by their work, to desire, achieve, or appreciate any intellectual attainments.[37] I.D.M.'s hierarchy is less rigid: he admits rudimentary education for the lower classes and even takes into account the possibility that an occasional brilliant student from amongst them would deserve to be accepted into the Academy. But basically the Academy admits only the sons and daughters of the aristocracy and of the rich bourgeoisie.[38]

In the egalitarian utopias selection for higher learning, however, is based solely on intellectual merits. Those selected for advanced studies are potentially the group of the future generation of magistrates. In other words, no matter how egalitarian, the utopists were realistic enough to know that not everyone is suitable for intellectual pursuits, that only a few are naturally gifted and inclined to such work. At the same time they believed, perhaps less realistically, that this minority of talented people, if properly trained, were the most qualified to rule and govern. Let us quote some accounts of this meritocratic system:

In the whole city and its neighbourhood, exemption from work is granted to hardly five hundred of the total men and women. . . . Among them . . . those whom the people, persuaded by the recommendation of the priests, have given perpetual freedom from labor through the secret vote of the syphogrants so that they may learn thoroughly the various branches of knowledge. But if any of these scholars falsifies the hopes entertained of him, he is reduced to the rank of workingman. On the other hand, not seldom does it happen that a craftsman so industriously employs his spare hours on learning and makes such progress by his diligence that he is relieved of his manual labor and advanced into the class of men of learning. It is out of this company of scholars that they choose ambassadors, priests, tranibors, and finally the governor himself. . . .[39]

---

[37] Patrizi, pp. 134-5.
[38] Antangil, p. 140.
[39] More, pp. 72-3.

The justification for the policy of electing the learned to govern is given by Campanella:

We, indeed, are more certain that such a very learned man has the knowledge of governing, than you who place ignorant persons in authority, and consider them suitable merely because they have sprung from rulers or have been chosen by a powerful faction. But our Hoh, a man really the most capable to rule, is for all that never cruel or wicked, nor a tyrant, inasmuch as he possesses so much wisdom.[40]

Andreae's Christianopolis is ruled by a triumvirate, with each of the three a personification of a virtue. The first is the chief priest 'from whose countenance there shone real divinity';[41] the second is the judge, 'a calm and peace-loving soul . . . the *pater familias* of the city';[42] and finally the director of learning (called Abida 'the father of knowledge' in Hebrew) — 'it was thought there was little he did not know'.[43] The triumvirate corresponds exactly to the three aims of education as presented by Andreae: 'Their first and highest exertion is to worship God with a pure and faithful soul; the second, to strive toward the best and most chaste morals; the third, to cultive the mental powers.'[44]

Wisdom, learning, and piety are also the characteristics of the elected rulers in the Republic of Eudaemon; in the Imaginary Republic the government is, in fact, a hierarchy of judicial magistrates — all versed in law and considered highly intelligent. Burton too expresses his preference for scholars:

Judges . . . and all other inferior Magistrates to be chosen as the *Literati* in China, or by those exact suffrages of the Venetians, and such again not to be eligible, or capable of magistracies, honour, offices, except they be sufficiently qualified for learning, manners, and that by the strict approbation of deputed examinators: first Scholars to take place, then Soldiers; for I am of Vegetius his opinion, a Scholar deserves better than a Soldier, because . . . a Soldier's work lasts for an age, a Scholar's for ever.[45]

Reverence for intellectuals is expressed by other means as well, such as exemption from work, preferential treatment in food, and the erection of statues in memory of those who contributed to knowledge. Some intellectuals, such as the professors of theology in Stiblinus's Republic or the priests and astrologers in the City of the Sun, although not wielding political power, are the most respected in the land and regarded as spiritual leaders.

The Academy in Antangil is the institution from which all the office-holders of the Kingdom must graduate. In the long chapter describing

---

[40] Campanella, p. 229.
[41] Andreae, p. 179.
[42] Ibid., p. 183.
[43] Ibid., pp. 186-7.
[44] Ibid., p. 209.
[45] Burton, p. 116.

its edifices, facilities, teachers, and the daily routine of the students, a detailed outline of the curriculum is also given. In the primary stage, from the age of six to the age of twelve, children are taught to read and write, grammar, poetry, history, music, and some elementary principles of geometry and cosmography. From twelve to eighteen the subjects are rhetoric, mathematics, logic, physics, and metaphysics, along with the study of the finest and most elegant orators and the best parts of medicine. In their spare time the students are taught designing, painting, architecture, fortification, and perspective. They continue with music and with exercises begun in the primary stage. They also engage in discussions to sharpen their minds. From eighteen to twenty-four, in addition to previous subjects, they learn the laws and ordinances of the Kingdom; they learn how to speak in public and in the law courts. The most religious and accomplished scholars may study theology and later enter the ministry. In all three stages they receive thorough military training and engage in mock battles with the young artisans of the city.[46] The Academy is, therefore, a primary and secondary school, a college and a university combined. Its purpose is both educational and political: to create the perfect ruling class on which depends the perfection of the entire system. I.D.M. follows in the footsteps of Thomas Elyot.

Not all utopists provide such a detailed account of the courses; but each of them wanted to express his own predilections. Thomas More ridicules casuistry, sophistry, and superstition through the programmes of higher education in Utopia:

In music, dialectic, arithmetic and geometry they have made almost the same discoveries as those predecessors of ours in the classic world. But while they measure up to the ancients in almost all other subjects, still they are far from being a match for the inventions of our modern logicians. In fact, they have discovered not even a single one of those very ingeniously devised rules about restrictions, amplifications, and suppositions which our own children everywhere learn in the *Small Logicals*. . . . They are most expert, however in the courses of the stars and the movements of the celestial bodies. . . . But of the agreement and discords of the planets . . . they do not even dream.[47]

And, in complete accordance with More's tastes, Greek literature and philosophy take precedence over anything written in Latin. Patrizi, more than any of the others, stresses those subjects which are directly relevant to public service: painting is useful for designing and for city planning, grammar is important for discussions of the laws and for composing letters, and even music has practical functions since it guides the soul, it either pacifies or incites.[48]

---

[46] *Antangil*, pp. 104-14; summarized in Negley and Patrick, op. cit., p. 309.
[47] More, pp. 90-1, 102.
[48] Patrizi, p. 141.

There is no clear distinction in these imaginary societies between the political leadership and what we would call nowadays the 'professions'. Law, medicine, and theology, are first taught as part of the general curriculum of higher learning, and later, in depth, to those who intend to become lawyers, doctors, or ministers. The 'professions' — which in certain cases include astrology — are part of the ruling élite in the Renaissance utopias.

The 'pre-scientific' utopias contribute little to the question of the *content* of higher education. They fit well within the current schools of thought and accept, basically, the existing curricula of the better schools. There is nothing peculiar to them or especially radical in their selection of subjects. Their contribution lies in extent and purpose rather than content.

We have one instance in our texts of another Renaissance figure — the Courtier — who is distinctly different from the Enlightened Ruler. In the Abbey of Thélème reside

People who are free, well-born, well-bred, and easy in honest company have a natural spur and instinct which drives them to virtuous deeds and deflects them from vice. . . . So nobly were they instructed that there was not a man or woman among them who could not read, write, sing, play musical instruments, speak five or six languages, and compose in them both verse and prose. Never were seen such worthy knights, so valiant, so nimble both on foot and horse; knights more vigorous, more agile, handier with all weapons than they were. Never seen ladies so good-looking, so dainty, less tiresome, more skilled with fingers and the needle, and in every free and honest womanly pursuit than they were. . . .[49]

These beautiful people have the facilities of 'the fine great libraries of Greek, Latin, Hebrew, French, Italian, and Spanish books . . . fine wide galleries, all painted with ancient feats of arms, histories, and views of the world'.[50] But no realistic utopist could afford to include in the perfectly efficient society such a group of human beings who are free from all social responsibility, who spend their days strolling in splendid surroundings and wear their learning as an ornament. It was Elyot's 'Governour' and not Castiglione's 'Courtier' who belonged in the utopian societies of the Renaissance. Thus, the Enlightened Ruler may resemble the Courtier in some skills and manners, but his education is far more utilitarian and prepares him, not for a life of courtly leisure, but for his heavy social functions; it was the ideal of the *vita activa politica*, which the utopists had inherited directly from the civic humanists of the preceding generations.

[49] Rabelais, p. 159.
[50] Ibid., p. 152.

(3) *The Scientist*

When . . . they explore the secrets of nature, they appear to themselves not only to get great pleasure of doing so but also to win the highest approbation of the Author and Maker of nature. They presume that, like all artificers, He has set forth the visible mechanism of the world as a spectacle for man, whom alone He has made capable of appreciating such a wonderful thing. Therefore He prefers a careful and diligent beholder and admirer of His work to one who like an unreasoning brute beast passes by so great and so wonderful a spectacle stupidly and stolidly.[51]

This paragraph in More's *Utopia* is the first plea in the history of utopian literature for scientific research; an apologetic and mild plea, not followed by any systematic suggestions as to how it was to be done, nor, apparently, with any impact on the theory and practice of education or on future utopias. For the rest of the century science is almost totally absent from ideal lands; the only subjects mentioned are the long-approved mathematics, astronomy, medicine, and the science of agriculture — and even those were to be studied from books based on classical theories.

A sudden drastic change is marked by the first major utopia of the seventeenth century — the City of the Sun. The new era was launched by an indictment of the old:

(The Solarians) take great pain in endeavouring to understand the construction of the world . . . the same argument cannot apply among you when you consider that man the most learned who knows most grammar, or logic, or of Aristotle or any other author. For such knowledge as this of yours much servile labour and memory work is required, so that a man is rendered unskilful; since he has contemplated nothing but words of books and has given his mind with useless results to the consideration of the dead signs of things. Hence he knows not in what way God rules the universe, nor the ways and customs of nature and the nations.[52]

In his footsteps soon followed Andreae:

If a person does not here listen to reason and look into the most minute elements of the macrocosm, they think that nothing has been proved. Unless you analyze matter by experiment, unless you improve the deficiences of knowledge by more capable instruments, you are worthless. Take my word for it — to such an extent do they prefer deeds to words . . . or, to be brief, here is practical science.[53]

This new approach to science in utopian literature culminates in *New Atlantis*, where the pursuit of science becomes the main object of the utopian society: 'The end of our foundation is the knowledge of causes, and secret motions of things; and the enlarging of the bounds of human empire, to the effecting of all things possible.'[54]

---

[51] More, p. 106.
[52] Campanella, pp. 261, 229-30.
[53] Andreae, pp. 154-5.
[54] Bacon, p. 288.

Although some dimensions of the idea of progress are still missing,[55] although they still believed that knowledge is finite and can be collected all in one book,[56] these texts nevertheless are an expression of ardent belief in the importance of the scientific investigation of nature. Consequently, the content of the curriculum underwent radical changes. Chemistry, physics, anatomy, botany, as well as the older disciplines of astronomy, astrology, and mathematics, take precedence over grammar, rhetoric, and the Greek tragedies. Languages, both classical and modern, are dethroned: 'Not too much care is given to the cultivation of languages as they have a good number of interpreters.'[57] Libraries and schoolrooms are no longer sufficient for the needs of the new programme. Laboratories of all kinds, pharmacies, botanic gardens, and zoological collections are added, not to mention the caves, towers, and artificial lakes of Salomon's House. The necessity of specialization becomes apparent, although admittedly only reluctantly: the supreme rulers — Hoh, Abida, and the Fathers of Salomon's House — possess all knowledge, but the rest of the scientists are already specialized and have to co-operate with one another.

The old and the new intermingle in these texts. The Enlightened Ruler and the Useful Citizen are still in evidence fulfilling the same functions and receiving the same moral, vocational and military training. It is the content and the extent of academic education which has undergone a transformation. Higher education would no longer be reserved for the minority that consists of the future rulers and administrators, but extended to all children who continue to study together until they are ready to specialize. The level of education of the entire society would be raised thus to a much higher level. This, undoubtedly, is a much more 'utopian' dream than the relatively modest aspirations of the 'pre-scientific' utopists.

The purpose of the scientific education is not made very clear. The list of inventions produced by Salomon's House is indeed staggering, and one could see how they might ameliorate human life. But before Bacon the number of fictional inventions in the utopias is minimal; incubators in Utopia, the art of flying mastered by the people of the City of the Sun, a complex sewage system in Christianopolis. It seems, then, that material gains are not the underlying motivation for the new curriculum, but, rather, the belief of these men that the possession of

[55] See above, p.6.
[56] Campanella, p. 220.
[57] Ibid., pp. 228-9.

such knowledge, revealing the secrets of God's creation, is important *per se* — even without side benefits. Science may not directly make the society better, but — according to these three — a good society is a community that encourages scientific research.

This radical shift in educational ideals, expressing the waning of humanism and the rise of the belief in science, is perhaps the clearest development in the Renaissance utopias. In social ideals, political theories, or religious orthodoxy, it is hardly possible to point out a chronological progression. But in this particular field the gradual evolvement of completely new attitudes is quite easily discernible. It is also in this field that our utopists had the most influence. The educational thought of Campanella, Andreae, and Bacon, directly and indirectly, influenced Comenius and his followers, inspired the radicals among the English Puritans in the 1640s, led to the writing of other utopias such as Plattes's *Macaria* and Samuel Gott's *Nova Solyma*, and prepared the ground for the foundation of the Royal Society.[58] But the relevant fact for this study is that they introduced a new persona to utopian literature, one which was totally absent previously and would wax bigger and stronger in the future, that of the Scientist who unravels the mysteries of the universe and attains power to control human destiny.

B. METHODS, SCHOOLS, TEACHERS

As in all other periods in the history of educational theory, in the Renaissance, too, arises the demand to make the process of learning easier. The methods of memorizing and reciting are deplored by many. 'Grammar and languages not to be taught by the tedious precepts ordinarily used, but by use, example, conversation, as travellers learn abroad, and nurses teach their children (so Lod. Vives thinks best, Commineus and others).'[59] In order to facilitate learning, many utopists insist that it should start with the vernacular rather than with Latin, on the principle of advancing from the easy and comprehensible to the more difficult:

They learn the various branches of knowledge in their native tongue. The latter is copious in vocabulary and pleasant to the ear and a very faithful exponent of thought.[60]

Here they see to it that what they read, they actually understand, and what they do not understand they translate into their native tongue.[61]

---

[58] For an extensive discussion of their influence see Charles Webster, *The Great Instauration*, London, 1975, *passim*.

[59] Burton, p. 112.

[60] More, p. 90.

[61] Andreae, p. 211.

Visual aids and concretion of concepts is another method advocated. In Stiblinus's republic moral and philosophical precepts are ornamentally written on walls within the city; in the Abbey of Thélème there are galleries of painting depicting the history of the world. It is only in the 'scientific' utopias, however, that this method was perfected. All knowledge is depicted on the circular walls of the City of the Sun and children are taught while walking round them: 'boys are accustomed to learn all the sciences without toil, and as if for pleasure'.[62] A similar system is used in Christianopolis, and Andreae explains why: 'Truly, is not recognition of things of the earth much easier if a competent demonstrator and illustrative material are at hand, and if there is some guide to the memory? For instruction enters altogether more easily through the eyes than through the ears'.[63] This was a contribution to the theory of education equivalent in importance to the inductive method advocated by Bacon for natural philosophy.[64]

For moral training most of the utopists favour the method of observing and emulating adults. The young sit amongst the adults, or serve them, during the communal meals, in order to learn manners, to listen to edifying talk, and to be questioned by their elders;[65] in the Happy City the entire programme of education for children aged five to seven consists of observing adults in their daily pursuits; statues of great men are also meant to inspire the young: 'pictures and statues of famous men, with their manly and ingenious deed, are to be seen everywhere, an incentive of no mean value to the youth for striving to imitate their virtue'.[66]

One instance of conditioning in utopian education deserves special attention, and that is the ingenious manner by which the young Utopians are taught to despise the traditional symbols of affluence: precious metals and precious stones are associated in their minds with babies' toys as well as with base objects such as chamber-pots and chains of slaves.[67] This is the one and only example in our texts of conditioning by psychological methods more sophisticated than the elementary system of reward and punishment.

[62] Campanella, p. 224.

[63] Andreae, pp. 220-1.

[64] Frances Yates regards Campanella's *City of the Sun* as 'an occult memory system' and 'the ultimate source of the new visual education' which directly influenced Andreae, and through him Comenius whose *Orbis pictus* (1st edn. 1685) was a primer for teaching children languages by means of pictures (see F. A. Yates, *The Art of Memory*, London, 1966, pp. 377-8).

[65] More, p. 79; Campanella, p. 232; Andreae, p. 202.

[66] Andreae, p. 202; *Antangil*, p. 96.

[67] More, pp. 151-7.

In the field of educational reform Andreae is undoubtedly the most important of all the utopists. He was not simply repeating fashionable ideas, but himself a leader of a new movement. One cannot help quoting long passages from his work on the subject of schools and teachers, since even today they have retained some of their forceful appeal:

I saw a school, roomy and beautiful beyond expectation, divided into eight lectures halls where the youths, the most valuable asset of the republic, are molded and trained to God, nature, reason, and public safety . . . all this is not after the infamous example of the world. For when the world seems to love her children most of all, she often shuts them up in some out of the way, unhealthy, and even dirty prison. . . . Here all is open, sunny and happy, so that with the sight of pictures, even, they attract the children, fashion the minds of the boys and girls, and advise the youth. They are not baked in summer nor frozen in winter; they are not disturbed by noise nor frightened because of loneliness. . . . Their instructors are not men from the dregs of human society nor such as are useless for other occupations, but the choice of all the citizens. . . . The teachers are well advanced in years, and they are especially remarkable for their pursuit of four virtues: dignity, integrity, activity, and generosity. They prefer to spur their charges on as free agents with kindness, courteous treatment, and a liberal discipline rather than with threats, blows, and like sterness.[68]

If not always in such elaborate and eloquent style, other utopists too plead for better schools and teachers. Stiblinus wants the school to be the true 'temple' of his island; the Academy in Antangil is one of the most splendid buildings in the Kingdom. All teachers, instructors, pedagogues in utopian lands — whether priests or laymen — are carefully selected for their wisdom, piety and virtue, since such a great responsibility can be left only in the hands of the best in the land.

C. CULTURE

The repression of all expressions of individuality in any utopian scheme is practically inevitable. It is the price paid for the desired stability. The denunications of this aspect of utopias have become commonplace. The uniformity, the censorship, the central control on all modes of expression, are an integral part of any serious utopia. The views as to what may become a source of danger, however, are liable to change. Thomas More, writing in 1516, although imposing rigorous prohibitions on travel, public gatherings, and the voicing of political ideas, as yet saw no danger in the newly-invented printing-machines. A century later, Andreae was to be quite aware of the potential ideological perils that lie in freedom of printing: 'For beyond the Holy Scriptures and

---

[68] Andreae, pp. 205-7. For the influence of Andreae on the educational philosophy of Comenius, see John Edward Sadler, *J. A. Comenius and the Concept of Universal Education*, London, 1966.

those books which instruct the youth and aid the devotion of the citizens, little printing is done. . . . Scattering literature which expresses doubt concerning God; which corrupts the morals or imposes upon man's mind is not permitted.'[69] In New Atlantis, the Fathers of Salomon's House carefully censor the publication of scientific discoveries: 'We have consultations, which of the inventions and experiences which we have discovered shall be published, and which not: and take all an oath of secrecy for the concealing of those which we think fit to keep secret; though some of those we do reveal sometimes to the State, and some not.'[70]

Another object of apprehension for many Renaissance planners of ideal societies was the theatre, not only because it had always been associated with frivolity, but also since it was probably the most influential medium for expressing moral judgements and social criticism. Stage-players are not allowed to set foot in Christianopolis; Patrizi and Stiblinus forbid the showing of comedies and satires; and in Evandria: 'They would not allow to present in the theatres or on stages comedies or tragedies or any spectacles which could introduce to the people profligate customs.'[71]

All decorations, paintings, and sculpture must have an educational value; art for art's sake is a foreign notion to the creators of perfect communities — 'Socialist Realism' is only the modern version of the subjection of arts to ideological purposes in all utopian projects.

Music is less strictly controlled and apparently feared less as a corrupting and subversive force. It is the only form of art mentioned in the sombre utopia of Agostini; in many of these lands they have musical accompaniment to the communal meals and to the rituals of religion; in the temple of Doni's city a hundred kinds of music are played simultaneously (!) during the weekly ceremony; in the more devout utopias stress is laid on the singing of hymns. Patrizi, however, did believe that certain types of music were dangerous while others were more suitable for making people gay and happy. Thus he decided that in his ideal world children would be taught only the Lombardic form of music. The expulsion of the poets from Plato's ideal republic has become a symbol for the authoritarian control over the arts and the suspicion towards all artists in utopia. The Renaissance visions of social perfection appear to be no different. Admittedly, a Poet Laureate is mentioned in More's

69 Andreae, p. 194.
70 Bacon, p. 297.
71 Andreae, p. 145; Patrizi, p. 140; Stiblinus, p. 95; Zuccolo, p. 55.

Utopia; and in the dialogue between Doni's mad-man and wise-man we find:

— 'were there poets?'

— 'Yes, but they had to make use of their hands in doing something else than verse', though in consideration for their talents they were given only light work and were allowed sufficient leisure to write.[72] But these are the only signs of literary activity of any sort. It seems unlikely that forms of art such as poetry, drama, and fictional prose would flourish under these rigorous regimes of control and censorship and — as M. L. Berneri points out[73] — certainly no utopias would be written there.

The pastimes of the citizens are a major concern for the planners. There should be no spare time for mischief. 'Sinful' pastimes such as drinking and gambling are most strictly forbidden. The utopists would encourage either sports and recreations that are conducive to good health, or games which stimulate the mind, such as versions of chess. Religious occupations take up much of the spare time of many utopians: they are obliged to attend prayers and listen to improving sermons very frequently.

Public celebrations are thought useful as means to increase fraternity and solidarity among the citizens, and also as a controlled outlet for the need for amusement: 'none shall be over-tired, but have their set times of recreations and holidays, *indulgere genio*, feasts and merry meetings . . . once a week to sing or dance (though not all at once).'[74]

Thomas More in his Utopia created a unique breed of human beings who are constantly 'cultivating the mind' and consider this the highest of pleasures. They flock, voluntarily, to public lectures given in the hours before daybreak (!), they attend *en masse* courses in Greek philosophy given by the European visitors, they read incessantly, they hold discussions and continue to educate themselves all their lives. (Their interests and choice of subject-matter, not surprisingly, coincide completely with More's tastes. One suspects that had More become the king of Utopia, he would have dictated the programme of further education to the last detail.) These eager students are the humanist's compromise between his intellectual ideals and the practical needs of the planned society.

Another conflict which the utopists had to resolve was between, on the one hand, the wish to maintain the isolation of the imaginary

---

[72] Doni, p. 13.

[73] M. L. Berneri, op. cit., p. 7.

[74] Burton, p. 118; also in Patrizi, p. 129; Stiblinus, p. 82.

country as a protection against the contaminating influences of the outside world and, on the other hand, the need of the perfect society to possess all the knowledge accumulated by all the nations. The dilemma was especially strong for the 'scientific' utopists who were well aware of the fact that science cannot flourish in seclusion. But even before them many of our visionaries wished to advocate the advantages of learning from other cultures.

And, just as they immediately at one meeting appropriated to themselves every good discovery of ours, so I suppose it will be long before we adopt anything that is better arranged with them than with us. This trait, I judge, is the chief reason why, though we are inferior to them neither in brains nor in resources, their commonwealth is more wisely governed and more happily flourishing than ours.[75]

Only it was soon obvious that one could not rely solely on the information brought by chance visitors whose ship was wrecked on the utopian shores. The new solution was to be the use of secret agents. The utopian country will guard most carefully its own achievements, but it will send spies to gather knowledge from abroad.

Some discreet men appointed to travel into all neighbour kingdoms by land, which shall observe what artificial inventions and good laws are in other countries.[76]

They continually send explorers and ambassadors over the whole earth, who learn thoroughly the customs, forces, rule and histories of the nations.[77]

In Antangil this method is practised only as regard new medicines and cures. But in New Atlantis this one-sided trade was developed to perfection: every twelve years a mission, the 'Merchants of Light', is sent out, whose errand was only to give us knowledge of the affairs and state of those countries to which they were designed; and especialy of the sciences, arts, manufactures, and inventions of all the world; and withal to bring unto us books, instruments, and patterns in every kind.[78]

These spies do not disclose their place of origin but 'colour themselves under the names of other nations'. Apart from these selected agents, travel abroad is completely prohibited in all utopias; and thus the contact with the outside world is also carefully regulated. The educational benefits of foreign travel, the 'grand tour', or pilgrimages to centres of learning, are all sacrificed for the sake of safety in seclusion and protection from all 'contamination' imported from outside.

The Renaissance and the Reformation produced an impressive array of educational theorists: Vittorino da Feltre, Vives, Ramus, Melanchton,

[75] More, p. 56.
[76] Burton, p. 115.
[77] Campanella, p. 224.
[78] Bacon, pp. 277-8.

Johann Sturm, Sadoleto, the Jesuit Fathers, Roger Ascham, Comenius — to name a few — had formulated the doctrines concerning the importance of education from the earliest age, the need for serious change in the education of girls, the choice of masters, site, and equipment for schools, interest in the psychology of the child, hatred for scholastic books and methods, inclusion of modern subjects in the curriculum. The utopists, with the exception of Andreae perhaps, contributed few original ideas to the body of contemporary educational theory. As in other fields, their contribution lies in arranging these ideas within the context of an entirely reformed society and in illustrating, through the vivid depiction of a living–working community, how the prevalent educational ideals of the period can be put into practice. They emphasized the immense power of education to determine the behaviour of the population and to mould the community; they equated the level of learning and training with the quality of the entire society; they presented ideal types of citizens and of leaders together with a description of the methods to create them; they viewed the educational system as a whole rather than focus their attention only on certain specific aspects. The conclusion that they wanted the reader to arrive at is that, for the interests of the whole, the central authority must take upon itself the entire responsibility for providing the right education for its population. They brought to the fore the values of universal and free education, given equally to men and women, designed to train them to fulfil their roles in the social structure.

The utopian texts are also valuable to the historian of education as a reflection of the changes in cultural ideals. While Thomas More represents the intellectual aspirations of Erasmian humanism in depicting a cultural atmosphere designed to attract the European intelligentsia, Eberlin von Günzburg, only five years later, already appeals to the religious fervour that possessed Europe in the immediate aftermath of the Reformation. The utopian vision is no longer that of life according to reason but of life according to the Evangelical Law. The second sharp turn occurs between the grimly pious Counter-Reformationary ideals of Agostini and the scientific spirit of Campanella or, to put it schematically, from the Law of God to the Law of Nature. Needless to say, the delineation is not, in fact, so sharply focused — in every new stage there is always a residue of the old. But the shift in emphasis, the growth in prominence of new ideal types, the intangible change of atmosphere, are all clearly discernible.

Another aspect of utopian literature which emerges from the analysis of the educational and cultural ideals is that the genre imposes certain rules and limitations on the desires of the writer: to make higher learning available to all citizens, although desirable, is impractical when one is contemplating all the necessities of a well-organized society; idle intellectuals, although very attractive, have no place in the busy beehive; pacifism and the redundancy of military training, although the most wonderful of dreams, are an impossibility when one describes just one island of perfection in a sea of evil. Certain proposals in these utopias are thus a result of the conflict between the true ideals of the writer and the impositions of the genre that he chose in order to express them.

Finally, a word about indoctrination and conditioning. In the dystopia of Aldous Huxley, *Brave New World*, the reader is presented with the horrifying prospect of an endless number of tube-bred identical multiple twins, brainwashed in their sleep to believe absolutely in all the dogmas of the ideology. This, according to the author, is the logical outcome of all utopian schemes. The motto of the book is Berdiaeff's wish that no utopia should ever be realized. Certain points in the Renaissance utopias do indicate that at least some of their authors indeed desired absolute uniformity and as little individuality as possible. A faint foretaste of Huxley's multiple twins is to be found in the way Campanella plans to reproduce closely resembling people by eugenic methods based on astrology: 'the majority, since they are conceived under the same constellation, are of the same age and resemble one another in strength, manner, and appearance. This gives rise to much lasting concord in the state, for they treat each other lovingly and helpfully'.[79] Undoubtedly the dictatorial nature of the control over all forms of free expression could justify the accusation that all utopian schemes must lead inevitably to totalitarianism. But these fears, based on the experience of the twentieth century, ought not to distort the evaluation of these texts as historical documents. Moreover, after analysing their content, one arrives at the conclusion that these men were primarily concerned with improving society by raising the level of education, and not with the creation of mindless robots. The programme of education in these utopias is, in fact, no more a plan of indoctrination than is any system of education and moral training in the most open of societies.

---

[79] Campanella, in Negley and Patrick, p. 331.

4

PROSPERITY AND EQUALITY

The utopias of the Renaissance had acquired their renown by the proposals for a complete reconstruction of society on communist and egalitarian principles. But a wider survey reveals that there are three distinct groups of Renaissance imaginary societies, of which only one consists of programmes for the abolition of private property and of class divisions. A second group includes those visions of an ideal community where classes and property remain, yet they aim at a levelling of fortunes and the blunting of social strife. The third group is of utopias concerned solely with the interests of certain élites, in total disregard of the social and economic problems of the rest of the population. Thus, before analysing the socio-economic content of the Renaissance utopias, it is necessary to place each one in its proper category to determine the principles on which it was founded.

Francesco Patrizi believed in the inherent inequality of men. His ideal city is divided into two social strata: the 'citizens' and the labourers. The first group consists of the guardians, the magistrates, and the priests. The entire structure is constructed for their benefit: to create for them the ideal conditions that will allow them to reach the 'celestial waters' of wisdom and happiness; they have all the rights and privileges, and no duties but to lead a life of contemplation and virtue. Complete equality in possessions and in dignities prevents envy and ensures fraternity in the ranks of the élite. The rest of the population — the peasants, the artisans, and the merchants — exist solely in order to serve the élite. Patrizi uses various similes to describe their role in society; servants at a wedding feast, the builders of the roads and the coaches, and the suppliers of provisions, for the passage of the citizens to the celestial waters; beasts of burden who are made so crude by their toil and misery that they would never be able to rise to a more exalted level of existence. Not only are they totally deprived of any rights or opportunities, but they are not even allowed to have their families and relatives with them so that they should neither be distracted from their work by emotions nor encouraged to rebellion.[1]

---

[1] Patrizi, pp. 125, 129, 134-6.

Patrizi followed the Platonic model in some respects. He too began not from an imaginary voyage but from a philosophical analysis of happiness and virtue, which led him to the entire social structure that would answer the needs of those people who were capable of scaling the steep mountain to happiness. The division into two classes, and the treatment of the masses as helots without rights, are also true to the Platonic tradition. Yet there are also certain significant differences; Patrizi, although categorizing his upper classes according to functions, is not really concerned with their duties to the State but only with their attainment of bliss; thus he does not impose communism upon them, nor frugality, because property and emotions are not considered as obstacles to the ultimate purpose. For him it is the producers, rather than the rulers, who must not be diverted from their labours by emotions of affection or of greed. Patrizi's utopia is not concerned with justice but with happiness, the happiness of a privileged minority.

The Abbey of Thélème was also built for the sake of a privileged élite: the inmates were an aristocracy of intellect and beauty, a group of Renaissance courtiers. They too are totally exempt from work or worry, and they are served by an army of faceless artisans and attendants who provide them with all the pleasures and the luxuries.

I.D.M., the unidentified author of *Antangil*, was not concerned with the happiness of one class but with the State as a whole; yet he believed that the well-being of a state depended on the quality of its upper classes. A carefully educated and dedicated aristocracy was his panacea for all social and political problems. Unlike the 'citizens' of the Happy City or the noble inmates of the Abbey of Thélème, the upper classes of Antangil are not only endowed with privileges but also burdened with well-defined duties and functions. Their excellent education ensures their devotion to the State; it also engenders such fraternity among them that 'they have amongst themselves what is almost a community of all possessions.'[2] (Here it is the love of friends that leads to equality amongst the educated classes of society; while in Patrizi's utopia the enforced equality leads to fraternity by eliminating envy.) At the same time the gulf between the two tiers of society is stressed and emphasized:

. . . public officials are elaborately dressed and attended. It is very becoming to a great prince and to everyone of authority to be honorably dressed and accompanied when he appears in public. For the luster of clothing and pompous following inspire inestimable respect and reverence . . .; people judge that such persons are full of virtue and merit since they bear its marks and signs.[3]

[2] *Antangil*, Eng. trans., in Negley and Patrick, p. 303.
[3] Ibid., p. 311.

The basic idea — the all-importance of the education of the upper classes — is Plato's; the model, however, is an idealized conception of the kingdom of England.[4]

In the utopian fragment, *New Atlantis*, the reader is presented with a new élite: the scientists. In an ideal state, Bacon claims, they ought to be given all the opportunities for research, so that eventually the fruits of their labours will create a world of affluence, longevity, and freedom from pain. There is no discussion of economic or social problems in this text; one can only discern the absence of certain features from Bacon's ideal world: no indication of an egalitarian structure; no frugality, but an emphasis on splendour and abundance which are the benefits of scientific research; no state welfare — the family seems to be the institution which is responsible for the needs of its members, the Feast of the Family being the occasion for dealing with such matters: 'There, if any of the family be distressed or decayed, order is taken for their relief, and competent means to live.'[5] Elsewhere Bacon did express opinions about economic issues, but not in his utopia. His ideas on the economic responsibilities of the state will be mentioned below; New Atlantis, however, must be dismissed, together with the Happy City, the Abbey of Thélème and Antangil,[6] from the following analysis, since the creation of these four utopias was motivated by concerns and ideals unrelated to the prosperity of a nation or the well-being of its members. One must turn to the other imaginary societies in order to find the reactions of utopists to the major social and economic issues of the day.

The communist utopias are at the other end of the scale in their attitude to economic issues and to social justice. This group consists of four utopists who believed that social perfection could be attained only if a total reorganization of economic relations would bring a decent standard of living for all and absolute equality. However, it should be noted that each of the four utopists had a different motivation for hi; revolutionary programme, and the form of communism in each of these ideal societies differed accordingly.

Thomas More's motives were decidedly economic and social. The following passages leave no doubt, in my mind, that he seriously believed that the abolition of money and of private property was the only way to attain social perfection:

[4] See above p. 26.

[5] Bacon, op. cit., p. 279.

[6] The author of *Antangil* devoted one brief section to welfare, and one paragraph to the revenues of the State, which will be mentioned below. But basically he belongs to the group of utopists which were concerned with an élite and disregarded economic matters.

. . . it appears to me that wherever you have private property and all men measure all things by cash values, there it is scarcely possible for a commonwealth to have justice or prosperity — unless you think justice exists where all the best things flow into the hands of the worst citizens or prosperity prevails where all is divided among very few — and even they are not altogether well off, while the rest are downright wretched . . . the one and only road to the general welfare lies in the maintenance of equality in all respects. I have my doubts that the latter could ever be preserved where the individual's possessions are his private property. When every man aims at absolute ownership of all the property he can get, be there never so great abundance of goods, it is all shared by a handful who leave the rest in poverty. It generally happens that the one class pre-eminently deserves the lot of the other, for the rich are greedy, unscrupulous, and useless, while the poor are well-behaved, simple, and by their industry more beneficial to the commonwealth than to themselves. I am fully persuaded that no just and even distribution of goods can be made and that no happiness can be found in human affairs unless private property is utterly abolished. While it lasts, there will always remain a heavy and inescapable burden of poverty and misfortunes for by far the greatest and by far the best part of mankind. . . . In Utopia, where nothing is private, they seriously concern themselves with public affairs . . . where everything belongs to everybody, no one doubts, provided only that the granaries are well filled, that the individual will lack for nothing for his private use. The reason is that the distribution of goods is not niggardly. In Utopia there is no poor man and no beggar. Though no man has everything, yet all are rich. For what can be greater riches for a man than to live with joyful and peaceful mind, free of all worries — not troubled about his food or harassed by the querulous demands of his wife or fearing poverty for his son or worrying about his daughter's dowry, but feeling secure about the livelihood and happiness of himself and his family . . .? Then take into account the fact that there is no less provision for those who are now helpless but once worked than for those who are still working.

. . . In Utopia all greed for money was entirely removed with the use of money. What a mass of troubles was then cut away! What a crop of crimes was then pulled up by the roots! Who does not know that fraud, theft, rapine, quarrels, disorders, brawls, seditions, murders, treasons, poisonings, which are avenged rather than restrained by daily executions, die out with the destruction of money? Who does not know that fear, anxiety, worries, toils, and sleepless nights will also perish at the same time as money? What is more, poverty, which money alone seemed to make poor, forthwith would itself dwindle and disappear if money were entirely done away with everywhere.[7]

These forceful arguments are based solely on considerations of social pragmatism, with an astonishing disregard for religious justifications. The counter-arguments, raised by 'More', in defence of private property, are intentionally weak and transparently specious.[8] It is, I believe, a total misunderstanding of the entire purpose of this work to suggest[9] that More did not really propose communism as a solution to the ills of

---

[7] More, pp. 52-3, 146-7, 149.

[8] '*Utopia* argues for the ideal of communism by the best test available: More has given to Raphael Hythloday all the good lines'', in R. C. Elliott, *The Shape of Utopia: Studies in a Literary Genre*, Chicago and London, 1970, p. 48.

[9] As do R. W. Chambers, H. W. Donner, C. S. Lewis, E. L. Surtz, and others.

European society; nor is it correct to claim that More's motivation was basically religious — to abolish the sin of Pride.[10] The case, I believe, is quite the opposite: pride and greed are presented as obstacles to the abolition of private property and not vice versa, as the following quotation clearly shows:

Nor does it occur to me to doubt that a man's regard for his own interests or the authority of Christ our Savior . . . would long ago have brought the whole world to adopt the laws of the Utopian commonwealth, had not one single monster, the chief and progenitor of all plagues, striven against it — I mean, Pride. Pride measures prosperity not by her own advantages but by others' disadvantages. . . . This serpent from hell entwines itself around the hearts of men and acts like the suckfish in preventing and hindering them from entering on a better way of life.[11]

Communism, claims More, is the one and only way to free humanity from the fear of want and to attain social justice. If he is sceptical it is only as regards the possibility that European society will ever be sensible enough to adopt such a system.

Utopia is a large island with fifty-four big towns, all governed by the same laws. Christianopolis is only one town of about 400 people. This difference in scope is in itself a difference in the quality of the communism. The atmosphere in Christianopolis is that of a large family: 'For what we are in our home, they are in their city, which they not undeservedly think a home.'[12] It is in spirit closer to existing brotherhoods (such as the Brethren of the Common Life) and millennial sects, than to the State communism designed by More.[13] The tone is also different: unlike More, with his pragmatic approach, Andreae constantly emphasizes the religious and ethical aspects of the communal way of life, and presents the moral virtues as their own rewards. The terminology he uses is also ethical and religious rather than social or economic:

To money is due public *corruption*. . . . No one has any money, nor is there any use for any private money. . . . And in this respect the inhabitants are especially *blessed* because no one can be superior to the other in the amount of riches owned. . . . It is considered *disgraceful* by all that one should take more rest and leisure than is allowed.[14]

In Utopia and in Christianopolis the family is the basic unit of society. Campanella, however, insists that it is inconsistent to abolish private property while retaining the family because the harmful passions and

---

[10] J. H. Hexter, *More's Utopia: The Biography of an Idea*, Princeton University Press, 1952, p. 80.

[11] More, p. 150.

[12] Andreae, p. 168.

[13] In other works Andreae drew models of secret societies and fraternities — *Christianae Societatis Imago* (1616/19), *Christiani Amoris Dextera Porrecta* (1620) — and *Christianopolis* is quite similar to them in spirit, though not in content and intention.

[14] Andreae, pp. 196, 161 (italics mine).

desires still remain: 'They say that all private property is acquired and improved for the reason that each one of us by himself has his own home and wife and children. From this self-love springs. . . . But when we have taken away self-love, there remains only love for the state.'[15] And in this lies the crux of Campanella's communism: it is not designed in order to achieve perfect social justice, nor for its moral advantages — communism, including the abolition of the family, is, according to Campanella, the only way to eliminate individualism and to obtain total devotion to the State. The whole is for him far more important than its components: '. . . the race is managed for the good of the commonwealth and not of private individuals'.[16] His motivation for the introduction of communism is exactly the same as Plato's; except that Campanella applies its rules to the entire population of the city-state and not to the rulers alone.

Doni too, when describing his communist society, eliminates all family relations. But his motives are completely different from those of the other three utopists. Doni desired simplicity. The system in his society is free of all complications: he repeats what More said, that the absence of private property and of money eradicates crimes of theft, robbery, embezzlement, gambling, litigation concerning property, etc.; he also adds that the extreme frugality prevents the envy of neighbouring countries and thus does away with war, together with the entire paraphernalia of soldiers, martial ceremonies, and chivalry; the total absence of family ties rids society from every kind of strife and painful feelings:

. . . they never knew whose son one was . . . and this prevented the pain over the death of a wife, relatives, fathers, mothers and sons, so that there was never a need to cry. . . .

There would be no abuse, no one would be dishonoured, relatives would not be offended, people would not be killed, men would not be murdered, there would be no daily disputes, the females would not be a source of infinite pains, there would be no tumults of weddings, no secret deceptions of husbands, no procurement, litigations for desertion, murders over dowries or any deceptive traps set by scoundrels. . . .

You see, when one died there were no testaments which cause life-long litigation. You see, a father had no fear that a son would ruin all the property nor that he would die of hunger. . . .[17]

[15] Campanella, p. 225.

[16] Campanella, p. 235.

[17] Doni, pp. 9, 10, 11, 14. Doni's insistence on simplicity conforms to a trend which Hiram Haydn (*The Counter-Renaissance*, N.Y., 1950) has defined as a second dialectic stage in the development of Renaissance thought. 'The praise of the humble and the simple is one of the most persistent features of the Counter-Renaissance', (p. 92). But, on the whole, our utopias, in their variety, do not bear out Haydn's division but rather confirm the more widely accepted view of the general eclecticism and heterogeneity of Renaissance culture.

Life in Doni's world will consist of the barest necessities and a few harmless amusements such as music and art. The superstructure of government becomes practically superfluous. All the troubles of this 'vale of tears' would cease to exist. How sincerely was Doni proposing his scheme? This question, I believe, is even harder to answer than it is in the case of More. He seems to be saying, as the title of the utopia indicates: yes, I know it is mad, but is it not tremendously attractive?

Social justice, a religiously moral life, the eradication of individualism, and simplicity, were the four motives for the foundation of communism in the Renaissance utopias; yet the consequences are similar in many ways: absolute equality, no trace of exploitation, and a society in which each member contributes according to his abilities and receives according to his needs. The State provides for the helpless, and most crimes caused by relations based on property are eliminated. In different words all four repeat the same formula: 'They are rich because they want nothing, poor because they possess nothing, and consequently they are not slaves to circumstances, but circumstances serve them.'[18]

This literary communism of the Renaissance is unique in many respects. Unlike Plato's, as mentioned above, it is applied to the entire population and thus creates a classless society, affecting not only the political framework but also the social and economic structue. It is different from the communism of religious sects which withdrew from the world, because the utopias were offering a national policy based on the communal ownership of the means of production (to use an anachronism) and not only on the equal distribution of goods.[19] Yet, as the Marxists were quick to realize, it is also fundamentally different from modern-day theory and practice of communism. First, since it is not seen as a result of an inevitable historical process or of a popular revolution, but as a system imposed entirely from above. Second, the Renaissance utopias strongly emphasize the abolition of money as a basic condition for the realization of their programme, either because of the negative connotations of 'money' in Christian ethics, or because of the desire to return to the simpler system of barter. By contrast, latter-day communism does not regard the disappearance of money as a necessary condition but, if at all, as a possible side-effect. Third, in the pre-industrial world frugality, the severe restriction of consumption, was an inseparable aspect of any imaginary, if realistic, plan for an equal distribution of goods that would allow a necessary minimum for each person.

[18] Campanella, p. 238; More, p. 149; Andreae, p. 156, 171; Doni, p. 9.
[19] As pointed out several times by Karl Kautsky in his writings on Thomas More.

The nineteenth-century visionaries felt that they could forgo such limitations and conceived of a world where all could equally share in the abundance created by industry. Thus, as a chapter in the history of communist thought, the intermediate nature of the Renaissance utopias is particularly evident.

In the communist utopias the bulk of the social-cum-economic problems are to be solved by the very introduction of communism. For piecemeal solutions to the specific economic issues of the day one must turn to the third group of our utopias. Eberlin von Günzburg, Stiblinus, Agostini, Burton, and Zuccolo did not strive towards a classless society and complete equality, but desired to ameliorate the quality of life for the entire population, to facilitate social mobility, and to blunt the social strife by making the nation prosperous, by imposing restrictions on the rich and providing welfare for the poor. They grappled with questions which at that time were receiving the attention of many European governments: tariffs, price and wage regulations, monopolies, encouragement of new industries, usury, employment and welfare, and the subordination of local authorities. Francis Bacon in 1597 summed up the expectations of contemporary reformers in matters of economic policy: 'The opening and well-balancing of trade; the cherishing of manufactures; the banishing of idlers; the repressing of waste and excess by sumptuary laws; the improving and husbanding of the soil; the regulating of price of things vendible; the moderating of taxes and tributes, and the like.'[20] Our utopists discussed all these as well as other matters in detail, but unlike other social critics, they took advantage of the imaginary framework which spared them the necessity to find means to surmount deep-rooted institutions and customs.

B. MOBILIZATION OF ALL RESOURCES

Idleness was one of the most frequently denounced social evils. It was attacked on religious grounds; it aroused fears of unrest and violence in the hearts of members of the established orders; it was considered to be wilful malingering on the part of the unemployed; it was understood to be the cause of poverty and led governments to enact harsh poor-laws. It was mostly associated with the lowest orders. The utopists saw idleness mainly in terms of economic waste, and they denounced it in all social classes. Manpower was the main resource of the pre-industrial economy;

---

[20] Bacon, 'Of Seditions and Troubles' (1625), in Vol. VI of *The Works of Francis Bacon*, edited by Spedding, Ellis, and Heath, London, 1858, p. 410.

to the unsophisticated 'economist' who sought to raise the standard of living for all the members of the society, it was obvious that a more equal distribution of the national wealth — although a necessary condition — would not suffice unless the wealth of the nation was increased considerably. In order to achieve this, all available hands had to be mobilized. In both the communist and the 'welfare' utopias everyone is forced to be productive, work is obligatory and honourable, and since all share in it, the burden is made lighter and more attractive. 'The chief and almost the only function of the syphogrants is to manage and provide that no one sit idle, but that each apply himself industriously to his trade, and yet that he be not wearied like a beast of burden with constant toil from early morning till late at night.'[21] Whether the planner was a Protestant or a Catholic, the attitude to work was the same: there is no room for drones in the planned society. These utopias are a far cry from the other type of visions of a perfect world — the 'Land of Cokaygne' type — where the land produces an abundance of all things without any human effort.

. . . consider how large a part of the population in other countries exists without working. First, there are almost all the women, who constitute half the whole; or, where women are busy, there as a rule the men are snoring in their stead. Besides, how great and how lazy is the crowd of priests and so called religious! Add to them all the rich, especially the masters of estates, who are commonly termed gentlemen and noblemen. Reckon with them their retainers — I mean, that whole rabble of good-for-nothing swashbucklers. Finally, join in the lusty and sturdy beggars who make some disease an excuse for idleness. You will certainly find far less numerous than you had supposed those whose labor produces all the articles that mortals require for daily use.[22]

In Naples there exist seventy thousand souls, and out of these scarcely ten or fifteen thousand do any work, and they are always lean from overwork and are getting weaker every day. The rest become a prey to idleness. . . .[23]

The analysis may have been grossly incorrect — it is quite obviously not true, for example, that in the pre-industrial society all women were unproductive — but the fact remains that all the utopists, as well as other observers, felt that a great number of people in the existing societies were not useful producers, and that the entire economic situation would change if they were all put to work. Let us consider their proposals for increasing the amount of productive labour.

It is only in the communist utopias that women are explicitly mentioned as a group that should be better utilized. They are released from most of the household chores and occupied in productive work. As

[21] More, p. 69.
[22] Ibid., pp. 71-2.
[23] Campanella, p. 237.

a rule, they are assigned the lighter jobs and the professions that were traditionally reserved for women. But on the whole it is clear that in these utopian lands women would be far more active in industry, farming, teaching, and nursing than in any of the existing societies.

. . . one (craft) is learned by each person, and not the men only, but the women too. The latter as the weaker sex have the lighter occupations and generally work wool and flax. . . .

Agricultures is the one pursuit which is common to all, both men and women, without exception.[24]

There are occupations, mechanical and theoretical, common to both men and women, with this difference that the occupations which require more hard work and walking a long distance, are practised by men. . . . But it is customary to choose women for milking the cows and for making cheese. In like manner, they go to the gardens near to the outskirts of the city both for collecting the plants and for cultivating them. In fact, all sedentary and stationary pursuits are practised by women, such as weaving, spinning, sewing, cutting the hair, shaving, dispensing medicines, and making all kinds of garments.[25]

The married women make use of the knowledge which they acquired while in college. . . .[26]

Only two suggestions concerning women would have seemed outrageous to contemporaries: that in Utopia 'the feminine sex is not debarred from the priesthood'; and that both in Utopia and in the City of the Sun women go to war together with the men.[27] In this respect too, then, the communist utopias are more radical, because in the non-communist ideal societies, although women's work is mentioned (as teachers or as nurses), there is no indication of an attempt to change the role of women in society, and only the male population is regarded as a reservoir of manpower.

Except for full-time students preparing themselves for their social functions, no man in the utopian lands is exempt from work. In the classless societies the problem would not arise, but even in the other utopias the upper classes are not allowed to be idle: in Wolfaria

the nobility will earn its living by farming. Every hamlet will have a nobleman who may have as many fields as he can cultivate with two ploughs. He will be the mayor of the village . . . all the mayors will administer justice and deal with complaints of the subjects. . . .

The mayor, bailiff, baronet, earl, prince and king, will not have any benefits from their office . . . they will be paid out of the public purse according to the importance of their work.[28]

[24] More, pp. 69, 68.
[25] Campanella, p. 231.
[26] Andreae, p. 260.
[27] More, pp, 140, 125; Campanella, p. 240.
[28] Eberlin, pp. 122-3.

In Eudaemon, lazy aristocrats lose their titles;[29] in Agostini's imaginary republic, the rule that each person must choose a profession and begin work at the age of fourteen applies to the entire population including the landowning aristocracy;[30] Burton, too, although rejecting complete equality, will not prevent plebeians from attaining honours and titles and at the same time he will demand that the nobles be able to support themselves:

I will have several orders, degrees of nobility and those hereditary, not rejecting younger brothers in the mean time, for they shall be sufficiently provided for by pensions or so qualified, brought up in some honest calling, they shall be able to live of themselves . . . he that by riot consumed his patrimony and ancient demesnes, shall forfeit his honours. As some dignities shall be hereditary, so some again by election or by gift. . . . For I hate those severe, unnatural, harsh, German, French and Venetian decrees, which exclude plebeians from honour.[31]

In Evandria the aristocracy contributes its share by fulfilling the tasks of government and leadership.[32] This abolition of a class of leisure entails also the disappearance of the large retinues of parasites and innumerable servants; and in addition public kitchens in the communist utopias and in Agostini's republic reduces even further the need for servants: 'Servingmen and servingwomen are a rare thing, nor very noticeable, except in the cases of those attending the sick, those in confinement, or babies. The husband and wife perform together the ordinary duties of the home and the rest is taken care of in the public workshop.'[33]

The frugality of the utopian lands eliminates the waste of labour on the production of unnecessary luxuries. Thomas More explains: ' . . . in a society where we make money the standard of everything, it is necessary to practice many crafts which are quite vain and superfluous, ministering only to luxury and licentiousness',[34] while in Utopia there are no luxuries, clothes are made at home of long-lasting materials, and buildings are so well-maintained that hardly any labour is spent on erecting new houses.[35] In Wolfaria great care will be taken that no useless trades should exist, and that in the useful occupations there would be no more masters than labourers.[36] The 200 trades in Doni's starshaped city are also only those considered essential to the simple life

[29] Stiblinus, pp. 82-3.
[30] Agostini, p. 121.
[31] Burton, pp. 113-4.
[32] Zuccolo, pp. 49-50, 67.
[33] Andreae, p. 160.
[34] More, p. 72.
[35] More, pp. 69, 74.
[36] Eberlin, p. 126.

of the community; the artists in this world are not exempt from productive labour, but in consideration for their creative needs, they have the privilege of doing light work so as to leave them enough time and energy for painting, writing, or composing.[37] Thus, art can be sometimes practised as a hobby, never as a full-time occupation.

From an attack on the idleness of the rich and the superfluous trades which they encourage, the utopists proceed to extirpate idleness from among the lower orders. One of the worst problems which troubled contemporary social critics was the large numbers of beggars and vagrants, unemployed who lived in appalling conditions, surviving on charity, persecuted everywhere by the authorities.

For the communist utopists the problem of mendicancy was solved quite simply: he who does no work receives no food. 'He who was lazy and was forgiven once, twice and thrice, they ordered that he should not eat until he had done his labour.'[38] The other utopists repress begging by laws and severe punishments, and by forcing all able-bodied men to work:

I will suffer no beggars, rogues, vagabonds, or idle persons at all, that cannot give an account of their lives, how they maintain themselves . . . if able, they shall be enforced to work.[39]

It was forbidden to all to beg in public; with the right balance of honor and utility every one may enjoy the honest rest-periods.[40]

Zuccolo adopts the system of enclosing the idle poor in workhouses outside the city: 'The poor of the Province are kept a mile or a little less from the city in habitations which resemble monasteries, and they buy their needs at the expense of the public by working in different trades so that they should earn their living entirely or at least partially.'[41] And in Antangil the idle poor who refuse to work are sent to the galleys or to the mines.[42]

Prohibitions on begging and vagrancy, hospitals that serve as workhouses for the poor, and punishment by hard labour, were the common remedies wielded by European governments and city councils; except that their success in 'curing' the problem was very limited and superficial. The 'welfare' utopists imagined a world where these existing practices would be carried out far more efficiently, executed

---

[37] Doni, pp. 8, 13.
[38] Ibid., p. 11.
[39] Burton, p. 118.
[40] Agostini, p. 92.
[41] Zuccolo, p. 48.
[42] *Antangil*, p. 137.

solely by the state and not through private charities, and supplemented by additional measures of state welfare. They insisted for instance, that even invalids should be given some work, for their own good and for the benefit of the community:

So that in Evandria the blind, the lame and the deformed are also occupied in some task.[43] The lame serve as guards, watching with the eyes which they possess. The blind card wool with their hands, separating the down from the hairs, with which later they stuff the couches and sofas; those who are without the use of eyes and hands give their ears or their voice for the convenience of the state, and if one has only one sense, he uses it in the farms.[44]

The truly 'impotent' will be provided for by the state. And thus the utopian landscape was to be totally free from the masses of unemployed and unemployable poor who, in the eyes of contemporary observers, were darkening the horizons of Europe like locusts.

Travel, which is restricted in the utopian lands for reasons of health and ideology, is limited further for economic considerations and the fear of vagrancy. In Utopia a special permit is required in order to leave town and the traveller, wherever he goes, must work for his food. Wandering aimlessly without permission is severely punished:

Now you can see how nowhere is there any license to waste time, nowhere any pretext to evade work — no wine shop, no alehouse, no brothel anywhere, no opportunity for corruption, no secret meeting place. On the contrary, being under the eyes of all, people are bound either to be performing the usual labor or to be enjoying their leisure in a fashion not without decency.[45]

Pilgrimages will be greatly restricted by the authorities of Wolfaria: 'No one will go to the Saints, begging on the way. No one will go to the Saints even without begging, if he has no permission in writing from his priest and his bailiff. A rich man will pay ten florins for such a permission. A poor man will work for ten days, without pay, in order to receive his leave.' The number of holidays in this violently anti-Catholic utopia will also be drastically reduced, because 'God is angered by such pretence of goodness'.[46]

The monastic orders too were regarded by many Europeans as a group of parasites. In Utopia there are two sects of people who devote their lives to doing 'good works' and all kinds of menial tasks for the community and for individuals. They believe that in such a manner they please God.[47] Thus, the monastic orders of Utopia do not withdraw

---

[43] Zuccolo, p. 48.
[44] Campanella, p. 239.
[45] More, pp. 82-3.
[46] Eberlin, pp. 108-9.
[47] More, pp. 137-9.

from the world but, on the contrary, are more useful and productive than the other citizens. It goes without saying that no mendicant friar would be allowed to set foot in any of the utopian lands. Eberlin von Günzburg hated them with all the fervour of a renegade monk; he would have them all killed. There are no monastic orders at all in any of the other utopias except for the Imaginary Republic, and there it seems that the main function of the various orders present in the city is to take upon themselves missionary duties abroad. Thus, with the exception of Agostini's utopia, another group of non-producers has been eliminated from the busy beehive life of the utopian lands.

Thomas More was the first of the utopists to recommend penal servitude instead of imprisonment or execution of criminals. In book I of *Utopia* Raphael Hythlodaeus denounces the existing system of hanging thieves as immoral and self-defeating; he praises the system he discovered amongst the Polylerites in Persia, where criminals do hard labour, not in prisons or camps, but within the community. In Utopia itself criminals, as well as prisoners of war become slaves of the community and do the degrading tasks. The system of prisons is ridiculed since it means that the least deserving are fed freely at the expense of the public and do nothing to earn their bread.[48] Wolfaria followed Utopia: '. . . a thief will work as a common servant in the town and do all odd jobs; he will carry an iron chain on his feet . . . a robber will be a permanent servant of the community, as was said of thieves'; only murderers will be executed.[49] There are no prisons in the City of the Sun (hardly surprising in view of the long years of Campanells's life spent in the prisons of the Inquisition); prisoners-of-war are either sold or kept to do hard labour outside the city wall.[50] Most of the other utopists, however, avoided the issue and only I.D.M. and Burton mention in passing the existence of prisons in their imaginary states.

Wars were always an economic disaster, and maintaining armed forces was always an extremely heavy burden on the treasury and on the population in general. The utopists attempted to limit the detrimental effects of military necessities.

I hate wars if they be not *ad populi salutem*, upon urgent occasion. Offensive wars, except the cause be very just, I will not allow of. For I do highly magnify that saying of *Hannibal* to *Scipio*, in *Livy*: 'It had been a blessed thing for you and us, if God had given that mind to our predecessors that you had been content with Italy, we with Africa. For neither

[48] More, pp. 31-4, 112.
[49] Eberlin, pp. 128-9.
[50] Campanella, p. 247.

Sicily nor Sardinia are worth such costs and pains, so many fleets and armies or so many famous Captains' lives'.[51]

This seems to be the general attitude of the utopists: wars are sometimes inevitable (including some offensive wars for 'just' causes), but an effort should be made to minimize the extent of damage they cause in human suffering and in economic waste. In Wolfaria, for instance, farmers are not permitted to take up arms lest the normal production of food be interrupted.[52] A warning against causing unnecessary damage in wartime, such as burning, pillaging, killing of women and children, is repeated in several of the texts; preparations are made so that in case of war the burden on the population would not be too heavy: 'I will have . . . money, which is *nervus belli* still in a readinness, and a sufficient revenue . . . to avoid those heavy taxes and impositions, as well as to defray this charge of wars. . . .'[53]

Who will fight in these utopian wars? Most of the authors considered the maintenance of a regular standing army a waste and a danger.[54] In Christianopolis, the City of the Sun, Eudaemon, and Wolfaria, the citizens themselves are trained to defend their country; in Utopia the people are ready to fight if necessary, but they prefer to use mercenaries, whose lives they risk more willingly than their own;[55] in the Imaginary Republic and in Evandria there are professional soldiers, but they are the officers who lead the trained civilians into the battlefield; only Burton seems to rely entirely on a standing army. Thus, in general, one could say that the class of professional warriors, expensive to maintain, idle, and even harmful in peacetime, is also eliminated from the utopian scene.[56]

The fashion of duels began to be regarded as a dangerous vice towards the end of our period, and at least one of our utopists was already aware of the unnecessary waste of human lives involved in this practice. Zuccolo declares that: '. . . no one carries weapons in all Evandria, neither for beauty nor for pomp, nor to serve as a robber nor as a body-guard, as is customary among the Italians who bring ruin on their private homes and on the public treasury. But the Evandrians hardly understand the names Duel, Criminal Conspiracy or Feud.'[57]

---

51 Burton, pp. 121-2.
52 Eberlin, p. 126.
53 Burton, p. 122.
54 More, pp. 22-4, 46.
55 Ibid., p. 84.
56 See also above, p. 63 n. 30.
57 Zuccolo, p. 55.

In the non-communist utopias, military training and service is the only form of national duties required of the citizens. In the communist societies, however, the population is also recruited at times for public works; '. . . seeing that they are all busied with useful trades and are satisfied with fewer products from them, it even happens that when there is an abundance of all commodities, they sometimes take out a countless number of people to repair whatever public roads are in bad order. . . .'[58] In Campanella's city agricultural work is conducted on this basis — labour battalions going into the fields;[59] in Christianopolis 'There are also public duties, to which all citizens have obligations, such as watching, guarding, harvesting of grain and wine, working roads, erecting buildings, draining ground; also certain duties of assisting in the factories.'[60]

Supervision, legal compulsion, and social pressure to work are only one side of the coin. Besides the 'stick' there is the 'carrot'; and in the 'carrot', i.e. in the ideas for making work more attractive, lies the originality of our writers. In all utopian lands people will be well trained for their jobs, and therefore they will perform it better and with greater satisfaction. Andreae was particularly aware of the danger that Marx was later to name 'alienation', and he sought to prevent it in several ways, first, by education: 'The men are not driven to a work with which they are unfamiliar, like pack-animals to their task, but they have been trained long before in an accurate knowledge of scientific matters . . .'[61]; second, by encouraging a healthy competition among the artisans to give them simple pride in their handicraft; and, third, by fostering the feeling that they are their own masters producing for themselves, for their own 'home'.[62] In Utopia each person is allowed to learn more than one trade and to practise the one that he most prefers;[63] in the City of the Sun a person's trade is decided upon according to his natural inclination — no one is forced to do work for which he is not suited.[64] The burden is made lighter and no one is to work to exhaustion; the working-day is shortened to six or even four hours.[65]

[58] More, p. 75.
[59] Campanella, p. 248.
[60] Andreae, p. 168.
[61] Ibid., p. 154.
[62] Ibid., pp. 157-8.
[63] More, p. 69.
[64] Campanella, p. 245.

[65] See above, p. 42. This was less revolutionary, perhaps, in the sixteenth and seventeenth centuries than in the nineteenth century.

There are also various incentives and rewards: 'If any man deserves well in his office he shall be rewarded';[66] public honours are to be bestowed for exceptionally good service to the community and statues erected for distinguished contributions.[67]

However, the most important aspects of encouragement to work is the elimination of the stigma from manual labour. All these utopists — both Catholic and Protestant — repeat several times that, contrary to the world's opinion, work of any kind is honourable and idleness shameful:

... We shall declare that there is nothing ignoble in practising any of the arts which are not by nature despicable. . . .[68]

Pride they consider the most execrable vice . . . no one thinks it lowering to wait at table or to work in the kitchen or fields. . . . Every man who, when he is told off to work, does his duty, is considered very honourable. . . . The occupations which require the most labour, such as working in metals and buildings, are the most praiseworthy.[69]

But some qualification needs to be made. There are slaves in Utopia who do the slaughtering of livestock and other menial tasks which are considered unsuitable for the free citizens; in Wolfaria farming and metal forging are more honourable than other trades;[70] in Eudaemon too farming is considered to be a noble trade, though not so the mechanical trades;[71] in the Imaginary Republic and in Burton's utopia the aristocracy owns the land. Thus the link remained between nobility and land, and not all the utopists were able to free themselves from prejudices towards certain occupations. Nevertheless, altogether the utopists desired to blot out the scale of respectability of occupations and to accord dignity to every calling, so that each person would practise his trade with pride.

The utopists depicted cities, but their communities were basically agrarian, deriving their wealth — like the existing societies — from the land. Although placing the major emphasis on the mobilization of manpower, they did not neglect the importance of intensive and scientifically planned agriculture. '. . . they . . . make up for the defects of the land by diligent labor. Consequently, nowhere in the world is there a more plentiful supply of grain and cattle . . . the naturally barren soil is improved by art and industry. . . .'[72] Doni advocated specialization: in

[66] Burton, p. 117.
[67] More, p. 113.
[68] Agostini, p. 121.
[69] Campanella, pp. 237, 246.
[70] Eberlin, p. 126.
[71] Stiblinus, p. 116.
[72] More, p. 103.

each area they grow only one product, the one best suitable for the local conditions, and so the farmers become excellent experts in cultivating that particular crop.[73] All our writers insist that the available land should be utilized to the maximum:

. . . there is not a span of earth without cultivation.[74]

. . . there was not a foot of soil to be seen which was not under cultivation or in some way put to use for mankind.[75]

I will not have a barren acre in all my territories, not so much as the tops of the mountains: where nature fails it will be supplied by art.[76]

Burton goes on to describe the functions of special agricultural supervisors, whose object is to achieve maximum productivity:

. . . no depopulations, engrossings, alterations of wood, arable, but by the consent of some supervisors that shall be appointed for that purpose, to see what reformation ought to be had in all places, what is amiss, how to help it; Et quid quaeque ferat regio, et quid quaeque recuset, what ground is aptest for wood, what for cattle, gardens, orchards, fishponds, etc., with a charitable division in every village (not one domineering house to swallow up all, which is too common with us), what for lords, what for tenants; and because they shall be better encouraged to improve such lands they hold, manure, plant trees, drain, fence, etc., they shall have long leases, a known rent, and known fine, to free them from those intolerable exactions of tyrannizing landlords. . . .[77]

Similarly, in Agostini's Imaginary Republic, well-trained instructors go out to the villages to teach the farmers methods of cultivation.[78] Agostini is the one utopist who explicitly stipulated that sufficient fertile land was a *sine qua non* for the erection of a good society. Stiblinus expects the governors of the provinces in Eudaemon to be responsible for the proper cultivation of the land: if a government inspector finds neglect in the fields, the governor of that province is punished by dismissal, a heavy fine, and exile.[79]

These city-dwelling utopists knew little about agriculture and could offer little by way of significant reforms in methods of husbandry or the introduction of new crops. As with other specialized subjects, their suggestions — based more on literature than on first-hand experience — were therefore confined to general principles such as intensification of cultivation, planning and supervision, and instruction by experts, with a sprinkling of a few imaginative ideas as the wagons fitted with sails

---

[73] Doni, p. 7.
[74] Campanella, p. 248.
[75] Andreae, p. 143.
[76] Burton, p. 112.
[77] Ibid., p. 113.
[78] Agostini, p. 113.
[79] Stiblinus, pp. 117-18.

(Campanella), or the replanting of whole forests for convenience of transport (More).

The problem of enclosures troubled the minds of only two utopists — More and Burton — with a span of a century between them. More considered it to be one of the major causes for poverty in England, and he denounced it in the most vehement terms, coining the renowned phrase about 'sheep devouring men'.[80] From Burton' apologetic tone it is evident that the issue was still controversial, although by then the social problems involved in the enclosing of arable land for pasture must have been far less acute.

I will have no bogs, fens, marshes, vast woods, deserts, heaths, commons, but all inclosed (yet not depopulated, and therefore take heed you mistake me not), for that which is common, and every man's, is no man's; the richest countries are still inclosed, as Essex, Kent, with us, etc. Spain, Italy; and where inclosures are least in quantity, they are best husbanded as about Florence in Italy, Damascus in Syria etc., which are liker gardens than fields.[81]

The difference in attitude to enclosures was due no doubt to the time that had elapsed between the writing of these two utopias, but it also reflects the difference in motivation: More was concerned with social justice, Burton with national prosperity.

These two utopists — accustomed to the confinement of an island — were also the only two to deal with the possibility of over-population, i.e. that the land would not suffice for the needs of the rapidly growing numbers of people. They both suggest the solution of colonies.[82] For this More was denounced as the first instigator of British imperialism — a slight anachronism, to say the least.[83] Yet it is interesting to note that there were people in England, with its population of 4,000,000-5,000,000, who were aware of the population growth and believed that the island might soon become too crowded. More indeed seems to have been the first to raise such a possibility; later, towards the end of the sixteenth and the beginning of the seventeenth century, other voices were heard propagating colonization overseas as a solution to over-population, so that for Burton the idea was no longer a novelty.[84]

[80] More, pp. 24-7.
[81] Burton, p. 112.
[82] More, p. 76; Burton, p. 120. See also above pp. 35-6.
[83] Gerhard Ritter, The Corrupting Influence of Power, 1952. Shlomo Avineri, in War and Slavery, divides the interpretations of More's Utopia into several categories: humanist, Christian, socialist, neo-Catholic, and the 'German' power-approach, with its strong anti-English bias, which regards Utopia as a blueprint for British colonialism and imperialism.
[84] Robert Gray, 'A Good Speed to Virginia' (1609), in J. Thirsk and J. P. Cooper (eds.), 17th Century Economic Documents, OUP, 1972. In Plattes' Macaria (1641), the Council for New Plantations would organize state-aided emigration.

In most of the texts hardly a word is said about any natural resources except the land. More ascribes to Utopia the lack of sufficient iron ore, and it is the only item that they are forced to import. Evandria, we are told, is blessed with rich mines of gold and silver. But it was Andreae who, in accordance with his belief in the potentialities of science, was the first to insist that in an ideal state: '. . . everything that the earth contains in her bowels is subjected to the laws and instruments of science.'[85] In *New Atlantis*, of course, this becomes the central theme.

Increased production by the mobilization of all manpower and by the full use of natural resources is one facet of the means to ensure economic prosperity; restricted consumption is the other. Austerity was favoured by the utopists for several reasons: in order to curb pride and vanity, licentiousness and dissipation, which corrupt the moral standards of the nation; for health reasons, as indicated above; but most of all for economic reasons. Production according to the utopists' plans, will be considerably increased, sufficiently perhaps to allow for more leisure, for exports, and for reserves. Yet in the pre-industrial age it was inconceivable to think of limitless production. Therefore, in order to give every one a fair share of the national 'cake', it seemed essential that no one should consume too much. Also, both labour and bullion could not be wasted on the production or the import of luxury goods for the few when they were urgently needed to obtain essentials for all.

In the communist utopias the consequence is equal frugality for all: food, clothes, housing, are uniform and very modest. In the 'welfare' utopias severe restrictions limit the possibilities of spending money. Flaunting affluence is strictly forbidden; extra money can be spent on charity or for constructive purposes, never for ostentation. In these utopian lands there are no palatial mansions, no expensive ornaments or magnificent clothes, no dowries or extravagent funerals, no gargantuan feasts. Religious inclinations to asceticism and economic necessity combined here to form a rigorously austere ideal.[86]

The subject of external trade also received due consideration from our writers. Since they are wary of foreigners and of contacts with the outside world, the utopians, ideally, would seek complete self-sufficiency. But only Doni believed autarky to be feasible; the other utopists admitted the inevitable necessity of external trade, since no

---

[85] Andreae, p. 154.

[86] On the restriction of consumption: More, pp. 69, 73-4; Doni, *passim*; Campanella, pp. 235, 247; Andreae, pp. 152, 156, 169-71; Eberlin, pp. 126, 127, 129; Stiblinus, pp. 100-3; Zuccolo, pp. 51-2, 67.

country, however industrious or frugal the people, can produce all its needs, and every country needs markets for the surplus of its production. A more or less uniform policy of foreign trade emerges from these utopias. The prevalent belief was that the wealth of a country is measured by the reserves of bullion in her possession; therefore it must have a positive balance of trade: to import as little as possible and to export as much as possible. The reserves of gold and silver were to be kept for emergencies, mainly for wars.

... when they have made sufficient provision for themselves (which they do not consider complete until they have provided for two years to come, on account of the next year's uncertain crop), then they export into other countries, out of their surplus, a great quantity of grain, honey, wool, linen, timber, scarlet and purple dyestuffs, hides, wax, tallow, leather, as well as livestock. Of all these commodities they bestow the seventh part on the poor of the district and sell the rest at a moderate price. By this trade they bring into their country not only such articles as they lack themselves — but also a great quantity of silver and gold. This exchange has gone on day by day so long that now they have everywhere an abundance of these metals, more than would be believed. ... It is for that single purpose that they keep all the treasure they possess at home: to be their bulwark in extreme peril or in sudden emergency.[87]

This would have been the dream of every English mercantilist of the sixteenth and seventeenth centuries. The same principle is applied in the two remaining communist utopias where money has no value at home but is needed for external purposes. [88] Zuccolo is so anxious to retain as much of the bullion as possible that he decrees a law that is familiar in some countries even today: when going abroad a citizen of Evandria is allowed to take with him a very limited amount of money, the absolute minimum necessary for his travels. For the same reason foreign merchants who come to sell their goods at the ports of Evandria are not permitted to take their earnings back with them in the form of currency — they are forced to buy the products of Evandria (which are all of extraordinary quality) and thus reinvest the money.[89] The import of luxuries to any of the utopian countries is either prohibited (as in Wolfaria) or discouraged through high taxes:

Of such wares as are transported or brought in, if they be necessary, commodious, and such as nearly concern man's life, as corn, wood, coals, etc., and such provisions we cannot want, I will have little or no custom paid, no taxes; but for such things as are for pleasure, delight, or ornament, as wine, spice, tobacco, silk, velvet, cloth of gold, lace, jewels, etc., a greater impost.[90]

[87] More, pp. 83-4.
[88] Andreae, p. 195; Campanella, p. 247.
[89] Zuccolo, pp. 52-3.
[90] Burton, p. 115.

To protect local products, the utopian governments will not permit the import of goods that can be produced at home; the prices and quality of those things that are brought from abroad will be carefully regulated by the local authority.[91] In order to safeguard their isolation, but also because it was then considered sound economic policy, the utopians will endeavour to carry out all external trade themselves, in their own ships, by their own men. However, when they must deal with foreign merchants, they try not to let them enter the city or the province: 'they do business at the gates.'[92] The utopian trade policy is yet another example of the way in which the authors incorporated existing policies, or measures which other social thinkers were urging governments to adopt, into the framework of the imaginary societies where such policies would be carried out to perfection.

It is doubtful whether the utopists realized the extent of the planning required to achieve all the above goals. Very little is said about the matter. In Utopia we find that:

In the senate at Amaurotum (to which, as I said before, three are sent annually from every city), they first determine what commodity is in plenty in each particular place and again where on the island the crops have been meager. They at once fill up the scarcity of one place by the surplus of another. . . . When the time of harvest is at hand, the agricultural phylarchs inform the municipal officials what number of citizens they require to be sent. . . .'[93]

The 'supervisors' in Burton's utopia have similar planning functions; and in *Christianopolis* a few lines are devoted to industrial planning: 'The ones in charge of these duties are stationed in the smallest towers at the corners of the wall; they know ahead of time what is to be made, in what quantity and of what form, and then inform the mechanics of these items.'[94] The insistence on uniformity in economic matters: 'one currency, one coin and one economy',[95] and 'weights and measures the same throughout',[96] is also meant to facilitate planning and control by the central authorities on a national scale. But these are all the thoughts on planning that the utopists expressed — and one must admit that it is not an impressive collection. Admittedly, these men had no example in reality of economic planning in any modern sense; the policies of governments in those days were not based on a consistent, comprehensive programme but were a haphazard collection of measures adopted to solve

[91] Agostini, pp. 114, 117.
[92] Campanella, p. 247; Zuccolo, p. 52; Stiblinus, p. 108.
[93] More, pp. 83, 63.
[94] Andreae, pp. 160-1.
[95] Eberlin, p. 128.
[96] Burton, p. 121.

specific problems, that were then not even regarded as 'economic' but as political or religious. The utopists, rather than pin their hopes on better planning, relied heavily on supervision and the better enforcement of the same laws and regulations that had been enacted by existing governments. The utopia of the Renaissance is a 'planned' society in the sense that the original founder had built it upon a complete and detailed plan; but the office of the 'Planner' — so prominent in latter-day utopias — who constantly revises, corrects, lays down the programme of developments for the future, is almost completely absent from these extremely static utopian societies.

C. SOCIAL JUSTICE

With the abolition of money and private property in the communist utopias all inequality and exploitation will disappear automatically; the intensive productivity will increase the national wealth and ensure that the equal share for each person will be adequate; the state will be responsible for all those who cannot take care of themselves.

This universal behaviour must of necessity lead to an abundance of all commodities. Since the latter are distributed evenly among all, it follows, of course, that no one can be reduced to poverty or beggary.[97]

No beggar is known or tolerated; for they judge that if anyone is really in need, the republic ought not to have to be warned of its duty.[98]

We have seen that education and medical care are provided equally to all in both the communist and the non-communist utopias. It now only remains to be seen how the non-communist 'welfare' utopias proposed to relieve poverty and narrow the gap between rich and poor.

First, they reduced the number of poor by prohibiting the entrance of foreign vagabonds. Second, they eradicated unemployment by making work compulsory and providing training to everyone. However, the utopists, like the legislators of poor-laws, seemed to ignore the fact that many of the poor were unemployed not because they were shirking but because they could not find employment. Were the utopists implying that compulsion to work included the responsibility of the State to provide employment? If so, they did not say it explicitly. It seems more likely that it had not occurred to them that in the perfect society there might still emerge a situation of a surplus in manpower or any other form of undesired unemployment.

---

[97] More, p. 83.
[98] Andreae, p. 272.

For the truly 'impotent' poor, these utopias offer various measures of relieving them from their misery. There are some remnants of private charity: in Wolfaria and in Evandria the State imposes compulsory philanthropy on the well-to-do:

. . . on holidays everyone in church will give as much as he is able to afford for the poor.[99]

In Evandria you do not see a single person begging for alms nor do you find people who are too rich . . . and the magistrates encourage among the rich a constant competition in spending on public works. . . . So that simultaneously they take care of the comfort and decoration of the Province, the taste of the people and the income of the poor.[100]

In Antangil families are required by law to take proper care of their helpless relatives; but when a widow, an orphan, or an old person has no close relations to provide support, a tax is levied by the Centeniers from all residents of the parish, from each according to his wealth, and the money is given to the person in need — 'De ceste façon il ne trouve aucun pauvre.'[101] Agostini leaves some charitable functions to the Church: '. . . therefore I would want that this tithe should not flow into the barn of the rich pastor but that it would be divided by him or by the Bishop to those poor of his quarter who would be in more urgent need than others.'[102] But what is more unique about the utopias is the extent to which they put the onus on the governments — the above quotations have a sequel:

. . . what is needed above that money [the donations in church] will be given out of the public purse . . . Bailiff, law-court and all superiors will take great care of the poor. It will not be ordered by the priests. . . . All the poor who receive charity will wear badges.[103]

If they be impotent, lame, blind, and single, they shall be sufficiently maintained in several hospitals, built for that purpose; if married and infirm, past work, or by inevitable loss, or some such like misfortune, cast behind, by distribution of corn, house-rent free, annual pensions or money, they shall be relieved, and highly rewarded for their good service they have formerly done. . . .[104]

. . . They nominate certain *Elemosinieri*, persons respected because of their age, their conduct and their standing, who out of charitable motives go and spy out the needs of the poor, the infirm, the orphans, the old and the widows. And with the incomes of certain *monti* created for this purpose by the richest and most charitable people, they reduce their poverty by provisions and money, so that no one dies in Evandria disconsolate fearing that his young children would die of hunger.[105]

Welfare provided by the state can be meted out in different forms: distribution of food: '. . . the *Edili* distribute to the people of the lowest class flour, wine, oil and salted meat at the expense of the public

---

[99] Eberlin, p. 125. [100] Zuccolo, p. 67.

[101] *Antangil*, p. 137. This is I.D.M.'s one contribution to the problem of welfare, but it still leaves him in the group of elitist utopists. See above, pp. 80-1.

[102] Agostini, p. 120. [103] Eberlin, p. 125.

[104] Burton, p. 118. [105] Zuccolo, p. 65.

treasury. . . .'[106] Burton mentions free rent, and in Agostini's republic each citizen receives accommodation from the state in one of the uniform houses built by the government[107] — which seems to be one of the most progressive measures mentioned in the 'welfare' utopias.

Eberlin von Günzburg, whose utopia is to a large extent an expression of the grievances of the peasants, and many of its laws are to be found four years later in the Twelve Articles of the Peasants' War, insists on the rights of all citizens to benefit from the common property — to hunt, fish, and cut wood to the satisfaction of each person's needs. Burton's supervisors protect tenants from 'intolerable exactions of tyrannizing landlords' as well as from depopulations caused by enclosures.[108]

Old-age pensions paid by the State are mentioned by Burton as well as by Stiblinus.[109] Measures are taken to protect the poor from exploitation: prices are controlled by the government to prevent profiteering — this usually applies to the price of corn in times of dearth, but also to other commodities. In Wolfaria the law even prescribes the size of meals served in the taverns for a fixed price.[110] Control and regulation of wages are usually also included in the same context. Eberlin adds further details about the protection of the rights of journeymen: they cannot be dismissed without a cause, their wages must be paid promptly in cash on every pay-day, and a journeyman who falls ill must be kept and looked after by his master for two months. 'all *Fuggerei* will be abolished' in Wolfaria; the world of high finance and rich merchants who amass fortunes at the expense of the poor, personified for the Germans by the house of Fugger (which had been fiercely attacked by Luther, Eberlin's mentor), will not exist in the ideal land where all people earn their living by honest labour.[111] In the same vein, Burton objects to private monopolies since, he says, they 'enrich one man, and beggar a multitude'.[112]

Usury, the long-debated problematic issue, looms large in the 'welfare' utopias; and once again the difference in attitude is independent of religious affiliation. In Wolfaria the taking of interest is strictly forbidden.[113] Agostini, however, accepts basically the official attitude of the Catholic Church in that period: usury is fiercely denounced, but fair interest is permitted. Yet Agostini will not allow private persons to lend

---

[106] Ibid., p. 50. [107] Agostini, p. 85. [108] Eberlin, p. 125; Burton, pp. 112-13.
[109] Burton, p. 118; Stiblinus, p. 115. [110] Eberlin, p. 125; Stiblinus, p. 114; Agostini, p. 124; Zuccolo, p. 52.
[111] Eberlin, pp. 129, 124. [112] Burton, p. 121. [113] Eberlin, p.128.

money, and in order to prevent any such illegal transactions, he demands that the financial affairs of all the citizens should be continually recorded in the public registry. Money can be borrowed from the *Monte di Pietà*. The Franciscans had established this institution in several Italian cities during the second half of the fifteenth century; the Lateran Council of 1515 granted them permission to take a minimal rate of interest to cover their expenses, but thus it damaged its charitable functions. In the Imaginary Republic the *Monti di Pietà* is a government office which gives free loans to the poor but charges an interest of 3 to 4 per cent on loans to people with means.[114] Burton too adopts the idea of a public bank which enables the government to exercise control over the financial life:

Broker, takers of pawns, biting usurers, I will not admit; yet because we converse here with men not with gods, and for the hardness of men's hearts, I will tolerate some kind of usury. If we were honest, I confess, we should have no use of it, but being as it is, we must necessarily admit it. Howsoever most Divines contradict it, it must be winked at by Politicians. And yet some great Doctors approve of it, Calvin, Bucer, Zanchius, P. Martyr, because of so many grand lawyers, decrees of Emperors, Princes' statutes, customs of Commonwealths, Churches approbation, it is permitted, etc. I will therefore allow it. But to no private persons, nor to every man that will, to orphans only, maids, or such as by reason of their age, sex, education, ignorance of trading, know not otherwise how to employ it, and those so approved not to let it out apart, but to bring their money to a common bank (n. As those Lombards beyond seas, though with some reformation, mons pietatis, or bank of charity that lend money upon easy pawns, or take money upon adventures for men's lives) which shall be allowed in every city . . . at 5, 6, 7, not above 8 per centum, as the supervisors or *aerarii praefecti* shall think fit.[115]

Nothing better illustrates Burton's pragmatism — he will not attempt to create a New Man, but to accommodate ethical principles with men's inherent shortcomings. The statement that certain things in the political and economic life of a nation 'must be winked at by Politicians', sounds surprisingly cynical coming from the pen of a utopist and a clergyman; yet, in fact, it is quite typical to the realism of the Renaissance utopists in general and of the 'welfare' utopists in particular.[116]

Progressive taxation is mentioned explicitly only by Eberlin: 'A citizen who possesses less than 100 florins does not pay tax; but for every 100 florins one gives a farthing each week.'[117] The others may have taken taxation for granted; but they leave the reader perplexed, not only because of the absence of such an obvious measure for the levelling of fortunes but also because so little is said about the revenues of the State. It is only the author of *Antangil* (which does not belong in the category

---

[114] Agostini, pp. 124-6.   [115] Burton, pp. 120-1.

[116] On Burton's realism see William R. Mueller, *The Anatomy of Robert Burton's England*, University of California Press, 1952.

[117] Eberlin, p. 130.

of welfare utopias) who specifies how the expenses of government would be met. In Antangil the State owns the mines, land, woods, lakes, rivers, herds, and flocks. These sources of wealth are rented out and the rents are paid twice a year. If these funds are insufficient, everyone is taxed in proportion to the value of his property.[118] In the other utopias it remains unclear how they would be able to finance their welfare schemes.

Many of the issues discussed above were at the time on the agenda of most governments and municipal councils. The correlation between social harmony, economic prosperity, and military strength was becoming evident to many thinkers and policy-makers. Rulers and administrators were appropriating powers of interference in various spheres of economic life, encroaching upon the domains of traditional institutions and authorities. But it was as yet not only the pre-industrial age but also the pre-economic age: neither the legislators nor their critics formed a comprehensive policy; the measures adopted were formulated in political or religious terms; and were often inconsistent and contradictory. Moreover, the obligations of goverments to their subjects were not as yet fully grasped, if at all, and very few laws were motivated by a sense of duty to improve the standard of living of all the orders of society.

The utopian framework, on the other hand, lends the sum of the economic and social measures at least the semblance of a consistent and comprehensive theory. In the communist utopias the entire system revolves round the basic principle of communal ownership and the absence of money. In the 'welfare' utopias, although not founded on a single axiom, there are nevertheless a number of suppositions which constitute the basis for the economic structure. Agostini is probably the best example: his discussion of the utopian republic is divided into four parts — Sanità, Forma, Forza, and Richezza; in this last section he analyses the sources of national wealth which, according to him, are land, manufacture, and commerce.[119] He then proceeds to propose rules, both positive and negative, for the development and maximum expansion of each sphere. The same section includes also the measures for poor-relief and restriction of consumption because for Agostini, as for the other utopists, economic prosperity is significant only when it affects the entire population.

---

[118] *Antangil*, pp. 61-3.

[119] Agostini, pp. 111-14; unlike the eighteenth-century Physiocrats who regarded only the land as a source of national wealth.

Urban poverty was undoubtedly the severest problem of early modern Europe; the population growth in the sixteenth century aggravated it all the more. 'Poverty', writes Natalie Zemon Davis, 'was not usually shamefaced, did not remain quietly sick behind closed shutters; instead it poured into the streets with begging, noise, crime, threat of disease, and rioting.'[120] The authorities fought a constant, hopeless, battle against this greatest of all social evils by repressive poor-laws and by philanthropic institutions.[121] The major purpose of the punitive laws and the charities was to rid Europe of the dreadful eyesore of beggars. The link between squalid poverty and disease was evident to all (in Venice 'the Provveditori alla Sanità' assumed the chief responsibility for the enforcement of the poor-laws');[122] the threat to public order, municipal pride, as well as the influence of Humanists' writings (such as Juan Luis Vives's *De subentione pauperum*, 1532), led national and local authorities to make vigorous efforts to relieve the worst of poverty and to abolish begging.

The utopists, it is true, adopted many of the existing measures and anticipated some others: they too would punish idlers with all severity, enclose the unemployed in institutions resembling the hospitals of the period which served as workhouses, infirmaries, and centres of apprenticeship; they too would forbid usury and encourage charity. Yet the utopists aspired to more than just the 'cleansing' of the streets from beggars and vagrants — their aim was a more equitable distribution of wealth. Therefore they not only elevated the 'hospital' system to a national level, but also imposed heavy social responsibilities on the upper echelons of society; in dignifying labour, they made idleness into a vice not only of the poor but equally of the privileged classes; and even when they had no wish to undermine or revolutionize the social structure, they nevertheless aspired to far more equality and social mobility. Thus it is not only the typical combination of fear, aesthetic malaise, spiritual self-interest and paternalistic pity which drove our visionaries to describe their ideal states, but also a keen sense of justice.

---

[120] N. Z. Davis, 'Poor Relief, Humanism and Heresy', in *Society and Culture in Early Modern France*, London, 1975, p. 24.

[121] For a good summary of the theories and methods adopted by the 'paternal state' to combat poverty and idleness see Robert Jütte, 'Poor Relief and Social Discipline in Sixteenth-Century Europe', *European Studies Review*, Vol. 11, no. 1, 1981, pp. 25-52.

[122] Brian Pullan, *Rich and Poor in Renaissance Venice*, Oxford, 1971, p. 632.

5

LAW AND ORDER

Why should there be laws and law-enforcement in the perfect society?
Prima facie one would have thought that in a vision of an ideal land there
would be no need for limitations on human activities nor penalties for
breaching such limitations. There are indeed three types of imaginary
societies in which the central theme is the absence of laws. First, the
land-of-Cokaygne, the Poor Man's Heaven, which is not a description
of a community but of a paradise for the individual who seeks total
freedom from restraint, obligation, and hardship. Second, there are
utopian writings, depicting ideal primitive societies where life is reduced
to the barest necessities, rendering all institutions redundant. Such is the
community discovered by Bonifacio's Portuguese sailors: 'a big island
full of people, who live without Prince, without laws and without the
knowledge of letters'.[1] This is also Gonzalo's dream in the *Tempest* (ridiculed by his listeners); and similar ideals are to be encountered in tales of
pastoral Arcadias or in glorified accounts of primitive societies
discovered in the New World. In such simple societies there are no rulers
to exact obedience, no religion to impose moral codes, no institutions to
create regulations. Each person satisfies his own simple needs without
friction with his neighbours, without aspirations for higher things.
Doni's 'Wise and Mad World' is the one utopia, from among our texts,
which belongs to this category. There are no explicit laws in this
imaginary society, only a description of a life of the utmost simplicity.
Behaviour is regulated by basic necessities, not by artificial laws. The
complete absence of social institutions, including family and property,
eliminates all emotions and motivations for crime. If there would remain
a few individuals with perverse innate tendencies to cause destruction,
they would be promptly exterminated together with any other cases of
deformity or incurable disease.[2] When there are no codes of laws —
either civil or religious — then there can be no law-breaking; where
there is no love, there is no hatred or jealousy. One must work in order
to eat — and the rest takes care of itself.

[1] Bonifacio, p. 10.
[2] Doni, p. 13.

A third type of an imaginary society without laws is that of a select group.

All their life was regulated not by laws, statues, or rules, but according to their free will and pleasure. They rose from bed when they pleased, and drank, ate, worked, and slept when the fancy seized them. Nobody woke them; nobody compelled them either to eat or to drink, or to anything else whatever. . . . In their rules there was only one clause: Do What You Will.[3]

A community of people so learned or so pious by nature that they require no external rules of conduct — instinctively all their actions would be moral. The Abbey of Thélème is one example, Comenius's Paradise of the Heart is another: 'The true Christian requires no copious Laws . . . for all are taught by one and the same spirit.'[4] Such a vision is based on the assumption that certain people in this world are intrinsically good, so that by careful selection and with the aid of learning, a community could be formed of individuals who would always act according to reason (or to ethical values) without any guidelines artificially laid down by social institutions.

The absence of laws and of law-enforcement is the hallmark of utopian fantasy; but in the serious utopias too we can find the claim that, even though laws and penalties cannot be completely dispensed with, the merits of the system are such as to reduce to a minimum the rate of crime and of immoral behaviour. Thomas More attributed this to the absence of money and property: 'Who does not know that fraud, theft, rapine, quarrels, disorders, brawls, seditions, murder, treasons, poisonings, which are avenged rather than restrained by daily executions, die out with the destruction of money?'[5] Campanella added the abolition of the family to the causes for the reduction in crime;[6] Stiblinus believed that the modesty and simplicity of life in Eudaemon would lead to virtuous behaviour;[7] Andreae relied on the piety of his imaginary people;[8] while Zuccolo pegged his hopes on the superior education of the inhabitants of Evandria: 'education makes people easy and compliant, not hard or litigious . . .', as well as on their improved economic and social structures: 'It is poverty that generates thieves and luxury greed. . . .'[9]

[3] Rabelais, p. 159.

[4] John Amos Komensky, *The Labyrinth of the World and the Paradise of the Heart*, trans. Count Lutzow, London, 1950, pp. 233-4.

[5] More, p. 149.

[6] Campanella, pp. 226-7.

[7] Stiblinus, pp. 101-1.

[8] Andreae, p. 164.

[9] Zuccolo, pp. 57, 59.

However, the serious utopists of the Renaissance did not believe that it was possible to transform human nature completely, and since Man was basically weak or corrupt, therefore, even in the best of states some coercion and restriction would be necessary. In the words of Andreae: '. . . penalties — there is no use of these in a place that contains the sanctuary of God and a chosen state. . . . Yet it must be confessed that human flesh cannot be completely conquered anywhere.'[10] Society, according to our writers, can be reformed only through good laws and good institutions that will protect men from the evil within them. The serious utopia is, in fact, a complex legal network of such close mesh as to leave individuals with very little freedom of action. There is a certain ambivalence in their attitude to human nature: on the one hand they seem to believe that a few exceptional individuals are capable of legislating for and ruling a good society with reason and benevolence (often it is quite clear that the utopist regards himself as the suitable person for this role: 'I have built this city for myself, where I may exercise the dictatorship');[11] on the other hand they have no such confidence in the masses: 'they distrust, not without cause, the self-control of human-beings'.[12] Thus, while they impose few limitations on their enlightened rulers, they tend to elaborate paternalistic systems of coercion for the rest of the populace. This ambivalence is apparently characteristic of many utopian works from all periods,[13] but it is undoubtedly strongest in the imaginary societies of the Renaissance which — as was said in the introduction — constitute a watershed: on the one hand they could not have been created were it not for the fact that a group of lay intellectuals had begun to believe that there is hope for mankind to emerge from the vale of tears by its own powers, without divine intervention and without waiting for the afterlife; but, on the other hand, they were still deeply infused with Christian doctrines of the Fall and the inherent weaknesses of human nature.

When based on a suspicious attitude to human nature, the legal system becomes one of the corner-stones of utopia. But there were other reasons why it was an issue of great importance to the Renaissance utopists. The sovereign as the fountain-head of justice was an integral part of the medieval conception of society. It was only to be expected, then, that in blueprints for perfect states, which augmented and

[10] Andreae, p. 164.
[11] Andreae, p. 140; also Burton, p. 109.
[12] Andreae, p. 174.
[13] See L. T. Sargent, 'A Note on the Other Side of Human Nature in the Utopian Novel', *Political Theory*, Vol. 3, No. 1, Feb. 1975.

enlarged the responsibilities of the rulers towards their citizens, the perfect dispensation of justice would loom large. When the utopists put the onus of public health, education, and welfare on the state they were adding to it new spheres of responsibility, but in demanding justice they were only repeating an accepted notion. Moreover, the historical period under discussion is notorious for its litigiousness. The entire world of courts, law-suits and lawyers was well-known to very large sections of the population. The judicial system was nowhere in Europe a side issue; it was ever present in the daily life of most people. Its abuses were also quite familiar and much commented upon, a subject for satirical attacks and innumerable proposals for reform. Rabelais would dismiss it all from his haven:

> Enter not here lawyers insatiable,
> Ushers, lawyers, clerks, devourers of the people,
> Holders of office, scribes and pharisees,
> Ancient judges who tie up good citizens
> Like stray dogs with cord on their necks,
> Your reward is earned now, and it is the gibbet.
> So go and bray there. Here is done no violence,
> Such as in your courts sets men fighting lawsuits.
> Lawsuits and wrangling
> Set us not jangling;
> We come here for pleasure.
> But may your leisure
> Be filled up with tangling
> Lawsuits and wrangling.[14]

The Law constituted the largest body of secular knowledge and the bar was the leading secular profession. Three of our utopists were jurists (More, Agostini, and Bacon), others had studied law but did not become practising lawyers (Doni, Rabelais), and it is safe to assume that most of our other writers were also acquainted with contemporary philosophy of law and had some legal training in either civil or canon law as it was taught at the universities of their respective countries. Thus, the legal system is a subject that received a great deal of attention from the utopists and deserves close examination.

The critics of the legal system in Renaissance Europe all deplored its artificiality, rigidity, and formalism, and the complexities that required the highest degree of technical refinement, verbal subtlety, and specialized scholarship. It was felt that these were developments which had brought about a separation of legality from justice, and that cases were no longer judged on the basis of moral considerations but of

---

[14] Rabelais, p. 153; see also Erasmus, *Praise of Folly*, Chapter 51.

technicalities. All our utopists, irrespective of their country, denomination, or particular background, acknowledged the widespread discontent with the existing legal system and, except for those who rejected the entire subject and took refuge in fantasy, devoted much thought and space in their descriptions of imaginary societies to offering an alternative in the form of an ideal legal system.

Laws, the utopists say, are guidelines for conduct. Codes of laws that are too long, extremely complex, open to different interpretations, and written in a language that most people do not understand, defeat their purpose. In all the utopian lands there would be: 'few laws, but those severely kept, plainly put down and in the mother tongue, that every man may understand'.[15] Knowledge of the laws should never be confined to a few experts, but must be the property of each and every citizen: 'With the Utopians each man is expert in law';[16] in Wolfaria one cannot become a 'house-owning citizen without demonstrating knowledge of 'the customs and the common laws';[17] in the City of the Sun the laws are 'written upon a flat table, and hanging to the door of the temple' for all to read and memorize;[18] in Christianopolis two tablets bear the sum of their beliefs and rules of daily life;[19] and in Evandria the laws are: 'reduced into certain little verses in the form of hymns, which they learn by heart, even children and village women, and they understand them without any commentaries or glossaries'.[20] The insistence that the laws be in the vernacular so as to be comprehended by all is, of course, an issue specific to a period when in most countries legal procedure and literature were in Latin or, in the case of England, in law-French. Even King James I (probably under the influence of his Attorney-General, Francis Bacon) called, in 1609, for the abolition of the use of Norman French which enabled lawyers to keep people in ignorance.[21] Other examples can be cited from all other European countries.

The demand for few and simple laws, arising from the condemnation of the intricacy of the existing system and from the general utopian quest

---

[15] Burton, p. 114; the same formula in More, pp. 53, 114-15 and in Andreae, p. 246.
[16] More, p. 114.
[17] Eberlin, p. 127.
[18] Campanella, p. 256.
[19] Andreae, pp. 175-8.
[20] Zuccolo, p. 56.
[21] Charles Ogilvie, *The King's Government and the Common Law 1471-1641*, Oxford, 1958, pp. 131-2.

for simplicity, could also be justified by the age-old theory of Natural Law: the belief that with the aid of reason men were capable of discovering a set of laws corresponding to nature and therefore of universal validity, immutable and eternal. The existing laws, claims Thomas More, are an artificial creation of the ruling classes for their own benefit:

the rich every day extort a part of their daily allowance from the poor not only by private fraud but by public law . . . a conspiracy of the rich, who are aiming at their own interests under the name and title of the commonwealth.[22]

He describes in Book 1 of *Utopia* how kings, with the assistance of sycophant lawyers and judges, daily impose new laws to enrich themselves at the expense of their subjects. The existing legal systems are a fraud; only a code of laws based on nature as interpreted by the divine gift of reason is a true code: 'The Utopians define virtue as living according to nature since to this end we were created by God. That individual, they say, is following the guidance of nature when in desiring one thing and avoiding another, obeys the dictates of reason.'[23] The people of Eudaemon too lead a virtuous life, says Stiblinus, not out of fear of penalties but because they desire 'secundum naturam vivere'.[24] The people of the City of the Sun 'know only the natural law', and in this they 'closely approach Christianity, which adds to the laws of nature only the sacraments, which give aid in observing those laws'.[25] The most concrete example of the utopists' belief in natural law is Bonifacio's application of the laws of the bees to a human community. However, the only full exposition of a natural law theory in the utopias is to be found in the introduction to the Imaginary Republic. Agostini expounds the Schoolmen's version which attributed the natural law to divine origin. The perfect civil law, based on natural law precepts, was embodied in the codex of Justinian, while the *Corpus Iuris Canonici* was the perfect codification of the divine laws as interpreted by man's reason and confirmed by Revelation. The corruption of this world is due to the departure from the original divine laws of Justice.[26] The ideal republic that Agostini then

[22] More, p. 148. This statement obviously delighted the nineteenth century Marxists. They could not have been similarly happy with More's belief in the Law of Nature which is of divine origin and not a superstructure of the relations of production.

[23] More, pp. 92-3.

[24] Stiblinus, p. 80.

[25] Campanella, ap. Negley and Patrick, p. 345. Both More and Campanella attempted to depict a pagan society in the New World which resembles Christianity as closely as possible. Life according to Natural Law was the best formula they could use.

[26] Agostini, pp. 21-7.

proceeds to delineate is to be based on a positive law which corresponds to the perfect laws of God.[27]

Roman Law is mentioned in two other utopias: Stiblinus, like Agostini, regards the pure Roman Law, without later 'barbaric' interpretations, as the ideal system of laws based on the immutable and eternal standards of nature, and therefore he makes it the constitution of his imaginary state.[28] Andreae expresses his admiration for the wisdom of Roman Law by making it a subject of study in the college of Christianopolis, but he refrains from claiming for it an absolute value and does not rely on it for the laws of his republic.[29]

The truth is that the utopists, with the exception of Agostini, did not have a real interest in philosophy of law; they used the theory that was at hand to endow their legislation for an imaginary society with a seal of absolute validity. They contributed nothing orginal to existing theories of law which their contemporaries, like Suarez and Grotius, were elaborating at the time, and simply paid lip-service to certain concepts which were convenient to their purpose.

The laws of each utopian land are ascribed to the founder of that perfect state who, like the mythical Lycurgus, was a legislator of infinite wisdom.[30] His laws were valid for eternity, no additions or changes would ever be required. Therefore, there is no need for a legislative authority. Edicts, statutes, proclamations, or decrees that create new laws and change old customs are unknown in utopia, for any novelty or change implies an imperfection in the original system. Even such dramatic occurrences as the revelation that had brought Christianity to New Atlantis or the conversion of the population of Antangil by Byrachil, a disciple of St. Thomas, had no effect on the laws which had been set down several centuries previously. The first law of the Garamanti in Guevara's *Diall of Princes* prohibits any additions to the seven laws of the country: 'We ordaine, that our chyldren make no more lawes then we their fathers do leave unto them; for oftentimes new

---

[27] Agostini, a practising jurist, wrote often about the abuses of the existing legal system. In Book 1 of the dialogues of *L'Infinito* he dwells at length on legislation, court procedures, the ideal judge, the corruption of lawyers. Therefore, in the Imaginary Republic itself, which is part of Book II, he only touches briefly upon these matters. See Firpo, *Lo Stato Ideale*, pp. 288-9.

[28] Stiblinus, p. 97.

[29] Andreae, p. 247.

[30] An allusion to another historical legislator was seen in the date which More chose for the foundation of Utopia. The year 244 BC was the year of accession of King Agis IV of Sparta who had initiated a set of radical reforms. See R. J. Schoeck, 'More, Plutarch and King Agis: Spartan History and the Meaning of Utopia', in *Philological Quarterly*, Vol. 35 (1956).

lawes make them forget old customes.'³¹ In the republic of Eudaemon
the Senate has power to legislate (with the consent of the people), but it
uses its authority only on very rare occasions, and Stiblinus endlessly
fulminates against the introduction of laws which destroy the harmony
of the traditional order. Admittedly, Stiblinus is one of the more
'reactionary' utopists; however, the static nature of utopian law does
not necessarily imply conservatism or a lamentation of a glorious past.
What some of the utopists propose as an immutable set of laws is often
quite 'progressive' or revolutionary. It is the notion of progress that is
missing: the belief in a continuous process of development in human
affairs which makes changes in social structure and in the laws an
inevitable consequence, is totally foreign to our thinkers. Even in
Bacon's 'scientific' utopia, although it is an ardent expression of faith in
the advancement of learning, there is no indication that the laws of
society would require any change whatsoever.

The absence of a legislative authority is one consequence of the
reduction of the code to a few and absolute laws; the elimination of the
lawyers is another. If the laws are known to all, if they are plain,
unblurred, and not open to different interpretations, then no special
expertise is needed for putting forward a case in a court of law. Lawyers
become redundant. The existence of lawyers was, according to the
utopists, a product of the artificiality, complexity, and vagueness of the
legal code, and as a result a law-suit depended not on the merits of the
case but upon the skill of the representatives of the parties. In ancient
Athens the use of professional legal experts in court was prohibited, and
both plaintiff and defendant had to present their own case; but it was
well-known that often their speeches were written by professional
logographers, among them the most eminent speech-writers of the day.
So there too cases were decided on point of verbal subtlety rather than
truth and justice. That is precisely what the utopists desired to abolish:

They absolutely banish from their country all lawyers, who cleverly manipulate cases and
cunningly argue legal points. They consider it a good thing that every man should plead
his own cause and say the same to the judge as he would tell his counsel. Thus there is less
ambiguity and the truth is more easily elicited when a man, uncoached in deception by a
lawyer, conducts his own case and the judge skillfully weighs each statement and helps
untutored minds to defeat the false accusations of the crafty. To secure these advantages in
other countries is difficult, owing to the immense mass of extremely complicated laws.
But with the Utopians each man is expert in law. First, they have, as I said, very few laws

---

³¹ Antonio de Guevara, *The Diall of Princes*, London, 1557 (Facsimile, Da Capo Press,
Amsterdam, 1968), trans. Thomas North, fo. 47.

and, secondly, they regard the most obvious interpretation of the law as the most fair interpretation.[32]

The need for lawyers symbolized for the utopists the imperfections of the system, but lawyers were also hated for their own faults. In Renaissance Europe lawyers were the most frequently attacked profession; they had become an object of satire and ridicule as well as hostility. They were considered not only a product of the corrupt system, but themselves largely responsible for its corruption. The prototype of the lawyer in contemporary literature is crafty, avaricious, narrow-minded ('the most unlearned sort of learned men', said Erasmus), arrogant, and pompous, motivated solely by selfish interest, ignoring truth, justice, and brotherly love.[33] Lying, distorting the truth, and obscuring the issues — their critics says — is the essence of their profession. They enrich themselves at the expense of their clients and make every effort to prolong trials and to prevent a just and swift conclusion. They are called parasites, vultures, plagues, and pests of the Commonwealth, belonging to the same category as usurers. Undoubtedly many lawyers had become very rich and powerful. Their exorbitant fees prevented poor people from obtaining justice. (Governments made attempts to cure this evil by forcing lawyers to serve the poor gratis: Wolsey imposed such obligations on the lawyers of England; local authorities in Italy often retained lawyers and paid them a salary for service to the poor. Agostini, out of charitable motives, gave free legal aid to the poor of Pesaro.) Moreover, people resented the very prominent position of lawyers in politics and administration. William Bouwsma considers the criticisms of lawyers a tribute to their remarkable and disproportionate influence in the molding of European society in those centuries: 'The deliberate exclusion of lawyers from More's Utopia only suggests their central place in the real world for one who understood it well. And an oblique tribute of a similar kind may be discerned in Luther's frequent expressions of hostility to lawyers.'[34] Indeed, lawyers were well represented in parliaments and in city-councils; in city-states like Florence they belonged to the ruling oligarchy,[35] and everywhere they served as

[32] More, p. 114. The same ideas in Eberlin, p. 128; Campanella, p. 254; Andreae, p. 246; Burton, p. 116; Zuccolo, p. 56.

[33] In Book 1 of *Utopia* the lawyer who defends the hanging of thieves is a figures of fun; pp. 20-8. Doni, in *I Marmi*, relegates all men of law to Hell, where they will be punished by being forced to eat for eternity their own dull writings.

[34] William Bouwsma, 'Lawyers and Early Modern Culture', *American Historical Review*, Vol. 78, No. 2, Apr. 1973, p. 316.

[35] Lauro Martines, *Lawyers and Statecraft in Renaissance Florence*, Princeton University Press, 1976, p. 5, and *passim*.

advisers to kings and princes, so that a contemporary pamphlet asks in desperation 'whether good lawes can be ordeyned and made where evil men of lawe shall be chiefe of counsell'.[36] They had come to represent in the eyes of many contemporaries the deterioration of morality, growing secularism, pragmatism, and relativism, or, as Bouwsma puts it: 'the lawyer was an obvious scapegoat for the general guilt of a world in transition'.[37] For almost every profession the utopists had portrayed an ideal model: an ideal ruler, knight, doctor, priest, judge. But nowhere did they depict an ideal lawyer as a counterpart to his despicable prototype in their view of the existing world — as one cannot have an ideal type of a usurer, so, they thought, there cannot be an ideal lawyer.

Only in three of the utopian lands do lawyers figure in any capacity: in Antangil they retain their traditional role, but they receive a salary from the government rather than fees from their clients; in Burton's utopia too lawyers are salaried officials but even so they are discouraged from appearing in court;[38] in Christianopolis they have only academic posts, and the sole function of the notaries is to copy documents.[39]

Judges, on the other hand, have an important role in these ideal societies. Admittedly, the judiciary was also often criticized by contemporary observers: they were accused of being a servile bureaucracy, lacking integrity or any professional standards of honour; they were notorious for bribery and partiality: 'These two evils, favoritism and avarice, wherever they have settled in men's judgements, instantly destroy all justice, the strongest sinew of the Commonwealth.'[40] In Book 1 of *Utopia* Hythlodaeus gives a vivid account how corrupt and frightened judges serve the interest of the king rather than justice.[41] But, the figure of the judge, unlike that of the lawyer, has an ideal counterpart: the magistrate in utopia is honest, intelligent, learned, pious, and merciful, unafraid of any person in power:

a man born after such a pattern that he does to no one what he would not wish done to himself, and what he desires for himself, he tries to secure for all. Neither blood nor riches, which here amount to nothing, exalted him; but a calm and peace-loving soul. He does not make his responses, confined, as it were, and seated on a tripod; and a citizen does not tremble at his looks; but like the rising sun, he shines upon all and clears up everything.[42]

[36] Quoted in E. W. Ives, 'The Common Lawyers in Pre-Reformation England', *Transactions of the Royal Historical Society*, 1968, p. 154.

[37] Bouwsma, op. cit., p. 316.

[38] *Antangil*, pp. 56-7; Burton, p. 116.

[39] Andreae, p. 247.

[40] More, p. 115.

[41] Ibid., pp. 44-5.

[42] Andreae, p. 183; also *Antangil*, pp. 53-6; Agostini, Book 1.

Burton, as usual more sceptical than the others about the virtue even of the best people, insists that the judges, like all state officials, should be under state control, give a yearly account of their deeds, and 'if they misbehave themselves, they shall be deposed and accordingly punished'.[43]

It seems, however, that the utopists were less concerned with the personal characteristics or the training of the judges than with the problem of the accessibility to justice. The courts of Europe were incredibly overburdened; the centralization of justice had been carried to extreme limits; the sessions were very short — all of which had aggravated the problem of prolonged trials and delays of years in reaching conclusion of lawsuits. Therefore, in Burton's utopia: 'Judges and other officials shall be aptly disposed in each Province, Villages, Cities, as common arbitrators to hear causes and end all controversies.'[44] In other utopias judicial authority, at least in the lower instances, is delegated to persons other than professional judges. Heads of families in Utopia and in New Atlantis are to be responsible for the behaviour of all the other members and to impose discipline and obedience.[45] In the City of the Sun: 'Everyone is judged by the first master of his trade, and thus all the head artificers are judges.'[46] Similarly, in Agostini's Imaginary Republic the guilds have extensive judicial powers. In Antangil the chiefs of ten households (like the tithingmen) and of 100 households constitute the lower instances of the judiciary; but in this rigidly hierarchical kingdom a member of the aristocracy has the right to be judged by his peers:

If it be between Nobles and Commoners, especially since the Nobles are ordinarily superb and haughty, and would not want to be judged by people of low quality: the Noble would take three of his friends and the peasant his *Centenier* and two *Dizeniers*, and they will make peace between them if possible; and if not they will be brought to the Judges. . . .[47]

The highest instances of justice, in most cases, are the rulers of each of these countries. They serve as the last instance of appeal, they have the power to pass death sentences and to give pardon.[48] The rulers themselves are not above the law, although the utopists do not elaborate what constitutional safeguards make such statements a feasible policy.

[43] Burton, p. 117.
[44] Ibid., p. 116.
[45] More, pp. 77, 112; Bacon, p. 279.
[46] Campanella, p. 254.
[47] *Antangil*, p. 57.
[48] Campanella, p. 221; *Antangil*, p. 55.

Heavy penalties for slander are meant to discourage frivolous legal actions;[49] Burton adds other measures to effect a reduction in unnecessary litigation:

He that sues any man shall put in a pledge, which if it be proved he hath wrongfully sued his adversary, rashly or maliciously, he shall forfeit and lose (n. It is so in most free cities in Germany). Or else, before any suit begin, the plantiff shall have his complaint approved by a set delegacy to that purpose; if it be of moment, he shall be suffered as before to proceed, if otherwise, they shall determine it.[50]

Thus, the burden on the courts would considerably diminish since the crime rate and the range of causes for litigation would be reduced by the utopian social system; false accusations would be prevented; there would be more courts and more judges as part of the judicial authority would be spread to various persons in positions of responsibility; lawyers, who delight in delaying justice, would not exist. In addition, the utopists imposed strict time limits for legal procedures: 'No controversy to depend above a year, but without delays and further appeals to be speedly dispatched and finally concluded in the time allotted.'[51] Others demanded an even swifter conclusion:

there is no written statement of a case, which we commonly call a lawsuit. But the accusation and witnesses are produced in the presence of the judge and Power; the accused person makes his defence and he is immediately acquitted or condemned by the judge; and if he appeals to the triumvirate, on the third day he is dismissed through the mercy and clemency of Hoh, or receives the inviolable rigour of his sentence.[52]

In Antangil the sentence is also passed on the first day of the trial;[53] Zuccolo allows for a maximum of fifteen days for the trial, one week for the first appeal, and three days for the second and final appeal.[54]

Few other details of legal procedure interested the utopists to the same extent. Several of them expressed their abhorrence of the practice of extracting evidence or confessions by torture:

They do not believe that the word forced out and extricated by the violence of torture is a sufficient proof for verifying crimes. Because the innocent and the timid, who have little courage, succumb, but the strong and obdurate, although guilty, come out free.[55]

They do not use torture because they are of the opinion that it may just as much make the innocent tell lies as the guilty admit the truth.[56]

[49] Eberlin, p. 123; Stiblinus, p. 106; Campanella, p. 256.
[50] Burton, p. 116.
[51] Loc. cit.
[52] Campanella, pp. 254-5.
[53] *Antangil*, p. 57.
[54] Zuccolo, p. 56.
[55] *Antangil*, p. 58.
[56] Zuccolo, p. 56; also Campanella. p. 255 and Agostini, Book 1.

The others did not mention explicitly their rejection of torture, but it is safe to assume that such methods would not be exercised in any of the utopian lands.

Campanella would require the evidence of five witnesses for a first conviction of a transgressor, and two or three for a second felony. It would not be difficult to find so many witnesses, he says, 'since they always walk about and work in crowds'.[57] The last matter of procedure mentioned in the utopias is Burton's rule that, in order to ensure fairness, 'All causes shall be pleaded *suppresso nomine*, the parties' names concealed, if some circumstances do not otherwise require.'[58]

Such is then the utopists' image of the ideal court: easily accessible to all citizens, regardless of riches or influence; the most honest and intelligent persons sit on the bench; the entire procedure is carried on in the vernacular; the plaintiff and the defendant present their cases without the aid of crafty lawyers; torture is never used, and instead a sound testimony is required; and the conclusion is reached promptly without unnecessary delays.[59]

The utopias depict centralized states that do not tolerate independent entities within them. There is only one Law and one hierarchy of courts in these imaginary lands. The battle that the secular authorities have been waging in Europe for centuries against the autonomy and the privileges of the ecclesiastical powers was resolved in the utopias totally in favour of the State. The judicial authority of the Church, if not completely abolished, is severely curtailed. Thomas More, whose imaginary state was conceived on the eve of the Reformation, and who was destined to die in this same battle but for the cause of the Church rather than the King, left to the clergy some legal powers and prerogatives. Priests in Utopia have immunity from the law: 'To no other office in Utopia is more honor given, so much so that, even if they have commited any crime, they are subjected to no tribunal, but left only to God and themselves. . . . It is easier for them to observe this custom because their priests are very few and very carefully chosen'; and also judicial authority, mainly in the sphere of morals, but even there the ultimate power to put into effect the judgement of the ecclesiastical court is in the hands of the secular authorities:

[57] Campanella, p. 256.
[58] Burton, p. 116.
[59] Very similar reforms were proposed by Thomas Starkey in *A Dialogue between Reginald Pole and Thomas Lupset.*

[The priests] preside over divine worship, order religious rites, and are censors of morals. It is counted a great disgrace for a man to be summoned or rebuked by them as not being of upright life. It is their function to give advice and admonition, but to check and punish offenders belongs to the governor and the other civil officials. The priests, however, do exclude from divine services persons whom they find to be unusually bad. There is almost no punishment which is more dreaded: they incur very great disgrace and are tortured by a secret fear of religion. Even their bodies will not go scot-free. If they do not demonstrate to the priests their speedy repentance, they are seized and punished by the senate for their impiety.[60]

In the utopias that follow, until we reach the reflection of the ideals of the Counter-Reformation, the situation is more clear-cut: no immunity for the clergy, no sanctuary from the Law in holy places, no powers to punish deviation.

They watch over the Ecclesiastics in all that pertains to their administration and conduct and even condemn them to all kinds of punishments if they fall into vices. Because in this Kingdom the Church has no right to temporal justice as there was none to the Levites among the Jews or to the Bishops in the primitive Church, but all is brought to the secular Judges.[61]

Agostini's Imaginary Republic is the single exception: there the secular and the ecclesiastical judiciaries will coexist side by side, 'like Moses and Aaron',[62] as in the sanctified state of affairs in Catholic Europe.

Needless to say, the abolition of the judicial powers of the Church does not imply greater secularism. On the contrary: at least some of our utopias are theocracies, and in all of them the State has appropriated the responsibility to supervise the faith and the moral life of the inhabitants. There is no distinction between crime and sin; there is no sphere of private morality — every action, thought, or disposition of the utopian citizen is a concern of the State. The State wields religious devices to ensure what it considers the right behaviour: 'It is the duty of all the superior magistrates to pardon sins. Therefore, the whole state by secret confession, which we also use, tell their sins to the magistrates.'[63] Religious tenets can serve the purpose of obtaining civil obedience:

[Utopus] gave injunction that no one should fall so far below the dignity of human nature as to believe that souls likewise perish with the body. . . . Who can doubt that he will strive either to evade by craft the public laws of his country or break them by violence in order to serve his own private desires when he has nothing to fear but laws and no hope beyond the body?[64]

The State punishes not only deeds of commission but also omission to observe virtues:

[60] More, pp. 139-40.
[61] *Antangil*, p. 56; also Eberlin, pp. 111, 128.
[62] Agostini, p. 105.
[63] Campanella, p. 257; also in More, p. 143; Andreae, p. 256; Bacon, p. 279.
[64] More, pp. 134-5.

The definitions of all the virtues are also delineated here, and here is the tribunal, where the judges of all the virtues have their seat. The definition of a certain virtue is written under that column where the judges for the aforesaid virtue sit, and when a judge gives judgement he sits and speaks thus: O Son, thou has sinned against this sacred definition of beneficence, or of magnanimity, or of another virtue, as the case may be. And after the discussion the judge legally condemns him to the punishment for the crime of which he is accused — viz., for injury, for despondency, for pride, for ingratitude, for sloth, etc. But the sentences are certain and true correctives, savouring more of clemency than of actual punishment.[65]

Even in Agostini's Republic, where there is an ecclesiastical judiciary, civil magistrates — the *veditori estraordinari*, whose duties also include the supervision of trade and industry — are responsible for watching over the morals of the population.[66]

Marriage and sexual relations are the main issues of morality that received the attention of the utopists. All of them list the laws and penalties related to pre-marital and extra-marital sexual relations, and some mention homosexuality as well. The reasons they give for the strict control over these matters are not wholly based on religion or morality; sometimes they offer pragmatic social considerations: 'unless persons are carefully restrained from promiscuous intercourse, few will contract the tie of marriage, in which a whole life must be spent with one companion and all the troubles incidental to it must be patiently borne'.[67] And if it is not for the protection of the family unit, then it is for the perfection of the race as in the City of the Sun where they practise eugenics. In any case, there is not a single utopia where sexual matters are not regarded as a major concern of the State.[68]

The complete congruence of the sphere of the law with the sphere of morals and the identification of sin with crime is not a product of the Counter-Reformation, as Firpo claims,[69] but a typical trait of all visions of perfect communities, particularly in an era dominated by religion. If the Lutheran Andreae came to express admiration for Calvinist Geneva, it was precisely for this eradication of any demarcation line between morals and laws.

And, as in Calvin's Geneva, the citizens of utopia are under constant surveillance: of their families, of their colleagues, or their neighbours

---

65 Campanella, p. 256.

66 Agostini, p. 104.

67 More, p. 109.

68 Admittedly, there are hardly any societies that leave these matters outside the sphere of the law. See the modern-day debate as summarized in Basil Mitchell, *Law, Morality and Religion in a Secular Society*, London, 1967.

69 Firpo, *La Stato Ideale*, pp. 302-3.

who are called upon to inform on every misdemeanour, of the police (even of secret agents), and of the various magistrates:

> . . . being under the eyes of all, people are bound either to be performing the usual labor or to be enjoying their leisure in a fashion not without decency.[70]

> If one realizes that someone lives in debauchery and spends more than his fortune allows, this person shall be denounced upon oath to the superiors.[71]

> A secret magistrate . . . keeps guard that there should be no conspiracies against the liberty of the fatherland and authority of the laws.[72]

Education, indoctrination, understanding of the laws, social pressure, and constant surveillance are all designed to ensure the correct behaviour of the inhabitants of the utopian world; but if all these safeguards should fail, the utopists are ready with a list of penalties.

Capital punishment was rejected by one utopist only: the inhabitants of Christianopolis, because of their religious convictions, 'are always chary of spilling blood, they do not willingly agree upon the death sentence as a form of punishment. . . . For anyone can destroy a man, but only the best one can reform.'[73] In other utopian lands the death sentence is not infrequent. It is the customary verdict for murder,[74] for conspiracies against the State,[75] or even for a potential conspiracy that may develop from discussions of affairs of state by unauthorised persons outside the parliament,[76] for adultery or pre-marital intercourse,[77] and even for pettier 'crimes' such as bankruptcy,[78] drinking to a person,[79] or defying dress regulations by wearing make-up or high heels.[80] The methods of execution mentioned in the texts are hanging, drowning, and stoning. Would there be professional executioners in these perfect communities? Most utopists ignore the question (for it is certainly an odd occupation to have in an ideal society); only Campanella says explicitly that there would be no such men: 'No one is killed or stoned unless by the hands of the people. . . . For they have no executioners and lictors, lest the state should sink into ruin.'[81] The City of the Sun is in many respects the most authoritarian of the utopian states, and some of

---

[70] More, p. 83.
[71] Eberlin, p. 129.
[72] Zuccolo, pp. 71, 72.
[73] Andreae, p. 165.
[74] Eberlin, p. 129; Campanella, p. 255; Burton, p. 119; *Antangil*, p. 59; Zuccolo, p. 59.
[75] Stiblinus, p. 100; Campanella, p. 255.
[76] More, p. 67-8.
[77] More, p. 112; Eberlin, pp. 113, 123; Burton, p. 119.
[78] Burton, p. 119.
[79] Eberlin, p. 123.
[80] Campanella, p. 236.
[81] Ibid., p. 255.

its peculiarities are alarmingly reminiscent (for the modern-day reader) of practices in twentieth-century totalitarian regimes. For instance: 'Certain officers talk to and convince the accused man by means or arguments until he himself acquiesces in the sentence of death passed upon him, or else he does not die. . . .'[82] This unusual demand is, perhaps, the key to the basic notion underlying the death sentences in utopian lands: certain criminals are executed not only in order to rid society of its 'rotten branches' but also to restore the balance of perfection that had been harmed by the criminal act. The acquiescence of the criminal to the sentence, implying his admission of the enormity of evil he had caused and his reacceptance of the ethical code of the perfect community, emphasizes the return to flawless perfection.[83]

Executions, obviously, are also means to instil fear and to deter from crime. This object is underlined in Antangil, where the convict is brought by procession through the streets to the gallows in the centre of the town, and he is left hanging for twenty-four hours, 'afin de donner plus de crainte et terreur aux meschans'.[84]

On the other hand, although they do not abolish capital punishment for what they consider to be the severest crimes, the utopists express an aversion to the common European practice of hanging thieves and other offenders against property. The mass executions of beggars and vagabonds was the desperate reaction of the authorities to the worst social problem of the day. Thomas More was the first to expound at length the injustice and ineffectiveness of death penalties for petty larcenies: unjust because no piece of property is worth a man's life, but even more so because many of those accused of theft are victims of a society that offers them no other way of earning a living — thus by hanging thieves the community punishes the wretches it had created by its own imperfect social structure; ineffective because many people resort to stealing as the only possible way to get food, and no punishment can deter the truly desperate; punishing all offences with equal harshness only increases the number of murderers among the thieves.[85] Several other utopists adopted More's arguments, though none of them expressed it so well and with the same insight into the social conditions that force men to resort to crime. 'The Evandrians are of the opinion

[82] Campanella, p. 236.

[83] Paul Foriers, 'Les Utopies et la Droit', in *Les Utopies à la Renaissance*, p. 241, compares this practice in Campanella's utopia to the Show Trials in Moscow in the 1930s.

[84] *Antangil*, p. 59.

[85] More, p. 29.

that one should not take the life of a person for simple theft, however large; because human life is far more precious than any treasure.'[86] Obviously though, even towards the end of our period, a century after More's fierce indictment of the system, the execution of thieves was still an issue that troubled the conscience of European humanists.

Another punishment intended to free society from its incorrigible offenders is exile or expulsion. For the utopians it meant an even severer penalty than for inhabitants of existing communities, since it involved not only being uprooted from one's homeland, but also being thrown out from the perfect haven into the world of evil; it meant that the person was branded as undeserving to belong to the community of the virtuous.[87]

A milder form of the same castigation is the ban — the excommunication of the offender not only by the religious authority but by the entire society:

Against the backsliders, especially those who remain stiffnecked after the vain warnings of brothers, fathers, and civil authorities, they pronounce the wrath of God, ban of the church, disgust of the state, and the abhorrence of every good man, with such success that it seems as if they have been shut off from the universe, that is all the creatures of God. They consider this more severe than death.[88]

The offender is completely isolated, ostracized, left in limbo. But it is less final than death or expulsion; by convincing repentance he may gain re-entry into society and a rehabilitation by the magistrates. Indeed, the utopists emphasize the need to allow penitent sinners to atone for their crime and return to society; irreversible punishments should be applied only to the severest of crimes; education is one of the main objects of penalties. 'Sins of frailty and ignorance are punished only with blaming, and with compulsory continuation as learners under the law and discipline of those sciences or arts against which they have sinned.'[89]

In addition to capital punishment, exile, and excommunication, the Renaissance utopias contain the whole gamut of penalties: hard labour (in the form of slavery for the community or for the injured party, as well as relegation to the mines or to the galleys) is the punishment for crimes such as vagrancy, adultery, theft, aggressive heretical propaganda, or the practice of alchemy; cutting off limbs or the tongue

[86] Zuccolo, p. 58; also Eberlin, p. 129; Burton, p. 119; Andreae, p. 165; and Thomas Starkey, *A Dialogue between Pole and Lupset* (London, 1948), p. 114: 'for every little theft a man is by and by hanged, without mercy or pity — which, meseemth, is against nature and humanity, specially when they steal for necessity without murder or manslaughter committed therein'.

[87] Stiblinus, p. 114; Campanella, p. 254; Andreae, p. 257; Zuccolo, p. 58; Bonifacio, p. 31.

[88] Andreae, p. 257; Stiblinus, p. 114; Campanella, pp. 227, 254.

[89] Campanella, pp. 255-6.

is applied in cases of slander, perjury, and blasphemy; flogging or public shaming; deprivation of certain rights such as marriage or drinking wine, and exclusion from the common table, from intercourse with women and from public prayers; and fines — in the societies that have not abolished money. Imprisonment, although not favoured by the utopists mainly for economic reasons, is also mentioned in Burton's utopia and in *Antangil*. In short, none of the known punishments is missing from the world of the ideal societies.

The utopists often claim that re-education is the principal object of penalties; yet, at the same time, they do not reject the principle of retribution: 'they pay an eye for an eye, a nose for nose, a tooth for a tooth, and so on, according to the law of retaliation'.[90] And one cannot avoid the impression that, despite some protestations to the contrary, the chief aim of the formidable list of penalties is to exact obedience by instilling fear. Patrizi admits this openly: 'But when I speak of the domestic enemy within the city, I say that the fear of punishment deters from execution of bad intentions.'[91] Those who do not obey the laws, who defy regulations or behave in any way differently from what is prescribed by the norms and customs, are the enemy within. They must be extirpated 'with iron'.[92] In the same manner that the utopian perfection justifies all methods of war with the enemies outside, so it hallows all ways of suppressing the dangers within the community of the blessed.[93] Law enforcement by fear (or terror, as we would call it today) seemed a perfectly legitimate weapon to the creators of the utopian societies. The strict regimentation and the harshness of the penalties create an atmosphere of grim military barracks rather than of idyllic happiness. The reliance on control, supervision, and coercion is incompatible with the repeated declarations of faith in education and good institutions. However, before we join the chorus of attack on utopias, let us take note of the following points:

(1) In comparison with the methods of law-enforcement in contemporary Europe, the proposals of the utopists were relatively lenient. They sincerely desired to reform the existing system of penalties out of genuine humanitarian motives. While in Europe vagabonds and beggars were 'everywhere executed . . . as many as twenty at a time on one

90 Campanella, p. 254.
91 Patrizi, p. 128.
92 Ibid., p. 129: 'col ferro bisogna troncarlo'.
93 See discussions on 'the paradox of perfection' in S. Avineri, op. cit.

gallows',[94] in the utopian lands no one would be killed for an offence against property. 'They punish most severely those misdeeds which are directed straight against God, less severely those which injure men, and lightest of all those which harm only property.'[95] Punishments in utopia are designed to fit the crime, and the laws prescribe the exact penalty for every offence so as not to leave room for arbitrary harshness (except in More's Utopia where the penalties are left to the discretion of the judges; and in Stiblinus's republic, where all offences are punished equally, since 'a trickle of small sins can easily grow into a river of crimes').[96] The utopists show an enlightened awareness of the fact that criminals are often victims of circumstances, and that society should reform itself rather than punish the products of its defects. They all abolish torture which in Europe was then a practice endowed with holy sanction by the methods of the Inquisition. Finally, they insist on mercy, pardon, and a gate of repentance because, despite their ultimate reliance on fear, they believed nevertheless in the object of re-education: 'Poor indeed is that physician who is more ready to burn and to cut, than to cleanse and to revive.'[97]

(2) When the modern-day reader of the Renaissance utopias is shocked by what seems to him an unnecessarily harsh penalty for a petty offence, he ought not omit to take into account the realities of the period. 'What need is there to punish someone who feels a sudden urge to ramble in the country?' asks M. L. Berneri.[98] But when More was proposing that a person caught twice travelling without a permit should be punished with slavery, he was not trying to suppress harmless desires for pleasant strolls; what he had in mind, was, again, the terrible European affliction of vagrancy. And there are other such instances of utopian penalties behind which contemporaries would have understood the motives and would have agreed that the utopian remedy was preferable both to the malady and to the remedy applied by the existing authorities.

(3) The Renaissance utopias were written at least two centuries before Mill's *Essay on Liberty*. The restriction of the interference of the Law only to actions that are harmful, and the whole concept of the intrinsic value of human rights and human freedom, had no roots in the *weltanschauung* of the Renaissance. The accepted notion was then, as with

---

[94] More, p. 20. Those are the boastful words of the lawyer at Cardinal Morton's table that had started the discussion on Justice.

[95] Andreae, p. 165.

[96] Stiblinus, p. 107.

[97] Andreae, p. 165.

[98] M. L. Berneri, *Journey Through Utopia*, p. 84.

Plato, that where Truth is known, there is no need for freedom of thought or action.

(4) The Renaissance social reformers were blissfully ignorant of the experience of twentieth-century civilization with totalitarian regimes.[99]

All this is not an apologia for the Renaissance utopias, only an emphasis on the need to see them within their historical context and to avoid evaluation in terms of 'good' or 'bad' that is determined by the historical experience of the following centuries. Undoubtedly, the utopian states are authoritarian in the extreme. There is a glaring discrepancy between the utopists' professed faith in education, better institutions, and true justice that should lead to less crime and law-enforcement, and their description of the innumerable possible infringements of the utopian law; the proud declaration that they would have few and simple laws based on nature is incompatible with the elaborate network of regulations and the extensive interference of the authorities in all spheres of life. The most important and highly valued characteristic of the utopian citizen is subservience: 'It is wonderful to see how men and women march together always obedient to their king. . . . They teach that the first concern must be for the life of the whole rather than the parts.'[100]

The authoritarian nature of the Renaissance utopias has been attributed to various causes: to several sources of influence that inspired the utopian vision of a perfect community, such as the Greek city-state (mainly Sparta);[101] the heritage of Plato;[102] the monastic communities;[103] Calvin's Geneva;[104] and even the Inca Empire.[105] Others regard the utopian quest for uniformity and order as a reaction to Renaissance individualism.[106] Many saw it as an inevitable consequence of the utopian

---

[99] Karl Popper admits in the Preface to the second edition of *The Open Society and Its Enemies*, that the virulence of his attack on Plato and his utopian successors is due partly to the fact that he wrote it during the second world war.

[100] Campanella, in Negley and Patrick, pp. 336-7.

[101] Lewis Mumford, 'Utopia, The City, and the Machine', in F. E. Manuel (ed.), *Utopias and Utopian Thought*; E. Rawson, *The Spartan Tradition*, pp. 170-85.

[102] Karl Popper, *The Open Society*, Vol. 1.

[103] Jean Séguy, 'Une Sociologie des Sociétés Imaginées: Monachisme et Utopie', *Annales*, Vol. 26, 1971; B. M. Bonansea, *Tommaso Campanella — Renaissance Pioneer of Modern Thought*, Washington D.C., 1969, pp. 395-6.

[104] Mainly in regards to Andreae, see introduction by Held to his translation of *Christianopolis*.

[105] 'It is very interesting to note that every single particular of Utopia mentioned by Mumford as inappropriate to an ideal commonwealth, is descriptive of Peru.' A. E. Morgan, *Nowhere was Somewhere*, p. 65.

[106] M. L. Berneri, op. cit., p. 56.

genre itself.[107] There is a probable grain of truth in each of these theories, but none of them is a full explanation by itself. The sixteenth and seventeenth centuries were a period when traditional organizations were losing ground and the national states were consolidating and extending their powers. Only the State, with its rapidly growing machinery of administration, could be regarded as capable of coping with the formidable volume of social and economic problems and with the fears of chaos. The ideology was wholly based on religious and ethical tenets, and examples of historical or contemporary communities were evaluated according to the pervasive ethos. The liberal alternative was non-existent. Natural Law theory supplied the seal of indisputable, absolute validity to ideologically accepted norms of behaviour. Those were the premises from which contemporary social visionaries inevitably began. They added to them the necessities of maintaining perfection in the face of Man's intrinsic corruption (of which they had no doubt), without the encouragement of a belief in progress. The end result was regimentation, coercion, encompassment of all human activities within the sphere of the Law, suppression of individuality, arrest of all change.

---

[107] To name but a few: Paul Bloomfield, *Imaginary Worlds or the Evolution of Utopia*, London, 1932; Gerhard Ritter, op. cit., J. D. Mackie, 'The Planner and the Planned For' in W. Nelson (ed.) *Twentieth Century Interpretations of Utopia*, 1968; as well as Popper, Avineri, Berneri, and L. Mumford, *The Story of Utopias*, 1962; J. L. Talmon, *Utopianism and Politics*, London, 1957, p. 14: 'Utopianism postulates free-expression by the individual and at the same time absolute social cohesion. This combination is possible only if all individuals agree. All individuals, however, do not agree. Therefore, if you expect unanimity, there is ultimately no escape from dictatorship.'

6

CONCLUSION

'Realistic utopias' appears to be a contradiction in terms, and yet the preceding chapters were an attempt to show that this was precisely the nature of the Renaissance descriptions of imaginary ideal societies. 'Realistic' because they were not dependent on any supernatural conditions or on any divine intervention which would change the cosmos, human nature, or the course of history. 'Realistic' also because they were not escapist dreams of fantasy or of science fiction — they remained well within the scientific and technological possibilities of their age; they grappled seriously with the major and the minor problems of their own society, offering practical, feasible solutions in minutest details. Many of their subjects of interest were so topical that they could not seek answers in classical sources but had to look for solutions in contemporary literature and practices. Moreover, contrary to a common notion based on a small selection of the Renaissance texts, most of these utopists did not propose a total and radical transformation of the social order but sought to alleviate human suffering and to ameliorate social conditions by improvements in existing institutions and practices. In other words, although imaginary by definition, these utopias were deeply rooted in actuality.

But they were unrealistic, or 'utopian', in a different sense: the societies of their imagination were drawn on a clean slate, built, as it were, on virgin soil free of all debris. Thus they avoided the necessity to indicate ways and means to change or overcome deep-rooted institutions and practices.[1] Therefore the utopias were not programmes for action. 'Utopia', writes Judith Shklar, 'was a model, an ideal pattern that invited contemplation and judgement but did not entail any other activity.'[2] The ideal would be realized only when a philosopher-king, a 'Utopus', would arrive suddenly to transform society with a magic wand. For this reason the Renaissance utopias cannot be regarded as

---

[1] The single exception is Eberlin's programme, in the first section of *Wolfaria*, for the drastic and gradual measures to eliminate the monastic orders.

[2] Judith Shklar, 'The Political Theory of Utopia: From Melancholy to Nostalgia' in Manuel (ed.), *Utopias and Utopian Thought*, p. 105.

tracts of political theory or a summons to revolution; and in this respect
the Marxists were right to dismiss them as mere wishful thinking. The
first utopia which included a programme for action appeared only
during the English Civil War (Winstanley's *The Law of Freedom*, 1652);
and after the French Revolution the classical type of model utopias died
out as belief in progress and in the potential of political actions led to the
emergency of ideologies which rendered obsolete the helplessness of
Renaissance intellectuals.

Karl Mannheim gives his own private definition to 'utopia' as the aim
of revolutionary movements which desire to overthrow the existing
system, as opposed to 'ideology' which he defines as the outlook of the
ruling classes who wish to preserve the established order.[3] However, the
Renaissance utopias were not in any sense the aims of revolutionary
movements, and it is thus not surprising that Mannheim chose to ignore
them and to pick out the figure of Thomas Müntzer as a suitable repre-
sentative of 'utopian' aspirations for the period under discussion. The
inability of our utopists to envisage means to attain the desired
perfection is the cause of More's sigh at the end of his *libellus*, 'that there
are very many features in the utopian commonwealth which it is easier
for me to wish for in our countries than to have any hope of seeing realized'.[4] For not only were they lacking a belief in the possibility to
transform society through political means, but they could not even
console themselves with a concept of gradual but inevitable progress.
Judith Shklar is right, I believe, in saying that the Renaissance utopia is
'nowhere' also in an historical sense: it is not in the past because the
writer was not harking back to a civilization that had existed before, and
it is not in the future since, unlike nineteenth-century visionaries, the
thinkers of the Renaissance did not believe that humanity would
constantly advance and improve. They were inclined to believe that Man
was capable of creating social perfection, but they could not tell him
how or when it was to be achieved.

However, a slight qualification needs to be made. Although our
utopists undoubtedly were not offering programmes for action, not all
of them were armchair intellectuals proclaiming their helplessness.
Eberlin, as we saw, was influenced by the revolutionary spirit of the
early stages of the Reformation. When he wrote *Wolfaria* he was not
without hope that one or the other of the German princes would accept
not only the appeal for a reform of the Church, but also the radical social

---

[3] Karl Mannheim, *Ideology and Utopia*, New York, 1936.
[4] More, p. 152.

programme which many saw implicit in Luther's early teachings. Thus, even if he did not join the zealots who set out to establish communities of saints, equally he did not share More's quiescence. Later the optimism aroused by the Lutheran revolution gradually died out. Eberlin's expectations of an imminent overthrow of the established order are no longer apparent in the utopian writings of his successors. But then, in the seventeenth century, come the group of utopists whose collective vision has recently been labelled 'Pansophia — the Dream of Science'.[5] Campanella, who certainly did not lack optimism and revolutionary zeal, took a leading part in the Calabrian revolt which was meant to hasten the imminent great transformation and made ceaseless efforts to persuade Pope and Kings to accept his vision and to help bring it about. Andreae too belonged to this category of European thinkers who were not satisfied to wait patiently for a great reformer, but tried to prepare the ground through the works of secret societies. And then, of course, Bacon who expressed in utopian form his ardent belief in the immense possibilities inherent in scientific research. In sum, then, it is not absolutely accurate to say that the Renaissance utopias are 'in *no sense* a declaration of historial hopefulness.'[6] It is true that, ultimately, they all pinned their hopes on an enlightened ruler and had no political programme, but not all saw themselves as completely helpless, and their view of the future was not entirely hopeless.

As serious blueprints for reform the utopias suffer from additional weaknesses, besides the absence of the bridge leading from reality to perfection. First, they had little to say on the subject of government. Therefore, neither are they in themselves a summons to political action, nor are they tracts of political theory. The views of their authors can be summed up in a few words: the ideal government is selective and meritocratic, and it possesses mainly executive and some judicial powers, but no legislative authority since the laws are immutable and eternal. These three elements — the ability to elect and to depose rulers, the assurance that those rulers would be the best products of an excellent education, and the fact that they are bound by good laws which they cannot alter — are the safeguards against tyranny. Not only to the modern reader, but even in comparison with the political thought of their own period, such theory appears to be too elementary and naïve.[7]

[5] By the Manuels in their *Utopian Thought in the Western World*, Part III. The term Pansophia was used by Comenius to denote all the different aspects of his hopes for 'a millennium based on calm and orderly science as a way to God' (ibid., p. 207).

[6] Shklar, op. cit., p. 104 (Shklar's emphasis).

[7] For the complexities and sophistication of early modern political theory see Q. Skinner, op. cit.

A second failing of the utopists arises from their inexperience. The utopists placed such heavy responsibilities on the State that they could be compared in many respects with the modern welfare state. But when reading them one is constantly aware that their authors did not even begin to fathom the extent of planning and the cost involved in such schemes. The matter of cost and state revenues could be avoided in the communist utopias where the state owns all property, but it remains an unanswered question for the non-communist societies; the thought devoted to the problem of planning is weak in both.

There is a further feature of the utopias which may cause disappointment to the reader who expects to find in them singular originality and imaginativeness. When analysing in detail the social and economic proposals of the utopists, we find that practically all their ideas were reflections of widely prevalent views, of observations of many social critics, and even of attempted reforms of local and central authorities. The utopists did not transcend the technological limitations of their age; they contributed little to the theory of the subjects which they discussed: medical science, educational and economic theory, or philosophy of law; and in each subject they were dilettanti rather than professionals. Their solutions depend almost entirely on the reorganization of existing systems and structures, without offering real changes in their content. Where organization could not be a remedy, they usually avoided the issue. They could not envisage a cure to the plague or any way in which to decrease the ignorance of the medical profession. On the whole (at least until the three major utopists of the seventeenth century), they accepted the curricula of the more advanced among the existing schools. They did not suggest new crops or innovations in methods of husbandry. Even their proposals for law reform were basically organizational.

It is immediately obvious that the utopian genre, reborn during the late Renaissance, expresses, even exemplifies, most of the values of civic humanism:[8] the secularization of political and historical thought 'free from the medieval entanglement with theological concerns'; the ethics of the *vita activa politica*; proposals for a programme of education which should prepare the citizens for active engagement in the affairs of their own age and state; looking upon antiquity not melancholically as a golden age never to be redeemed but as an exemplary model for the present. However, it is most important to note that the utopias form a distinct and separate group within the humanistic stream, and it is their

---

[8] As characterized by Hans Baron, op. cit., pp. 443-62.

uniqueness which the present study has attempted to underline. The strength and uniqueness of the utopists lie not in abstract theory nor in any particular details but in their *comprehensiveness*. Their aim was to ameliorate life in all its spheres, and they were aware, to an amazing extent, of the interdependence of all aspects of the social organism. With a singular insight into the complexities of the social organization, they presented a total picture of community life. In the words of Lewis Mumford: 'Utopian thinking . . . was the opposite of one-sidedness, partisanship, partiality, provinciality, specialism. He who practised the utopian method must view life synoptically and see it as an interrelated whole: not as as random mixture, but as an organic and increasingly organizable union of parts. . . .'[9] This special contribution of the utopists becomes all the more evident when one examines not just the few famous utopias but the group in its entirety, and in particular those 'minor' utopists who did not arouse the interest of scholars specializing in Renaissance philosophy. It is precisely their concern with the practical, down-to-earth, organizational details which sets them apart from the important theorists of the period and should mark them out for particular attention.

The totality of vision, the realism and the concern with the acute problems of contemporary society, are, to my mind, the special traits which make these texts such interesting documents for the historian. The emphasis of this study on the social content of the utopias was meant to fill a gap and to correct some distortions. First, the socialist interest in the imaginary societies of the Renaissance created the association between utopias and communist programmes — we have seen that the social visionaries of the sixteenth and early seventeenth centuries offered a varied selection of models for perfect societies of which only a minority were based on communist principles. Second, the presentation of the utopists as 'precursors' and 'forerunners' stressed the eternal and universal aspects of these writings severed from their historical context. Third, the fact that More and Campanella were appropriated by the historians of philosophy and political theory, tended to obfuscate the importance and the uniqueness of the entire group of these utopias as mirrors of the *social* complexities of their own society. A fourth distortion ought perhaps to be blamed on those who over-emphasized the influence of the New World on the utopian genre in the Renaissance and thus blurred the fact that it was their own country which they had in mind, the problems of their own society which they

[9] L. Mumford, *The Story of Utopias*, pp. 5-6.

attempted to solve. They were willing to borrow ideas from past and present models of social organization, but they modified and applied them to the needs of their own community. The real content of these works is, predominantly, their authors' reaction to their own social experience. The utopian solutions were determined, mostly, by contemporary necessities rather than by intellectual or personal psychological factors. 'L'état intime d'une societé' is what Lucien Febvre claimed to be the reward of the study of utopian thought of a specific period. And therefore, indeed, the genre that Thomas More had reinstated in European literature is a fruitful field of research not only for the social thinker, the historian of ideas, the sociologist, and the student of literature, but, not least, for the social historian as well.

BIBLIOGRAPHY

I. PRIMARY SOURCES

AGOSTINI, LUDOVICO, *La Repubblica Immaginaria*, ed. L. Firpo, Turin, 1957.

ALBERTI, LEON BATTISTA, *Della Architettura Libri Dieci*, Milan, 1853.

ANDREAE, JOHANN VALENTIN, 'A Modell of a Christian Society', and 'The Right Hand of Christian Love Offered'', in G. H. Turnbull, 'J. V. Andreae's Societas Christiana', *Zeitschrift für Deutsch Philologie*, 73 (1954) and 74 (1955).

—— 'REIPUBLICAE CHRISTIANOPOLITANAE DESCRIPTIO' IN F. E. HELD, *Johann Valentin Andreae's Christianopolis: An Ideal State of the Seventeenth Century*, OUP, 1916.

BACON, FRANCIS, *The Advancement of Learning and New Atlantis*, OUP, 1974.

BONIFACIO, GIOVANNI, *La Republica delli Api*, Venice, 1627.

BOORDE, ANDREW, *The Breviary of Helthe*, ed. F. J. Furnivall, EETS, London, 1870.

BUCER, MARTIN, *De Regno Christi*, translated by W. Pauck and P. Larkin, London, 1969.

BURTON, ROBERT, *The Anatomy of Melancholy*, London, 1926.

CAMPANELLA, TOMMASO, *La Città del Sole*, ed. N. Bobbio, Turin, 1941.

CASTIGLIONE, BALDASSARE, *The Book of the Courtier*, translated by Sir Thomas Hoby, London, 1974.

CROWLEY, ROBERT, 'The Way to Wealth', in *The Select Works of Robert Crowley*, ed. J. M. Cowper, EETS, London, 1872.

DONI, ANTON FRANCESCO, 'Il Mondo Savio e Pazzo', in C. Curcio (ed.), *Utopisti e riformatori sociali del cinquecento*, Bologna, 1941.

EBERLIN VON GÜNZBURG, JOHAN, *Ausgewählte Schriften*, Band I, Halle, 1896, in *Neudrucke Deutscher Litteraturwerke des XVI und XVII Jahrhunderds*, nos. 139-41.

ELYOT, THOMAS, *The Boke Named the Gouvernour*, ed. H. H. S. Croft, London, 1883.

ERASMUS, DESIDERIUS, *The Education of a Christian Prince*, translated by L. K. Born, New York, 1936.

—— *Praise of Folly*, translated by B. Radice, Penguin Books, 1971.

GUEVARA, ANTONIO DE, *The Diall of Princes*, translated by Thomas North, London, 1557. Facsimile by Da Capo Press, Amsterdam, New York, 1968.

HALL, JOSEPH, 'Mundus Alter et Idem', in *The Works of Joseph Hall*, Vol. 12, Oxford, 1837-9.

I.D.M., 'Histoire du Grand et Admirable Royaume d'Antangil', in F. Lachèvre, *La Première Utopie française: Le Royaume d'Antangil*, Paris, 1933.

KOMENSKY, JOHN AMOS, *The Labyrinth of the World and the Paradise of the Heart*, translated by Count Lutzow, London, 1950.

MORE, ST. THOMAS, *The Complete Works*, ed. J. H. Hexter and E. Surtz, Yale University press, 1965.

—— *Utopia*, ed. E. Surtz, S. J., Yale University Press, 1967.

PATRIZI DA CHERSO, FRANCESCO, 'La Città Felice' in C. Curcio (ed.), *Utopisti e riformatori sociali del cinquecento*, Bologna, 1941.·

PUCCI, FRANCESCO, 'Forma d'Una Republica Catholica', in D. Cantimori e E. Feist (eds.), *Per la storia degli eretici italiani del secolo XVI in Europa*, Rome, 1937.

RABELAIS, FRANÇOIS, *The Histories of Gargantua and Pantagruel*, translated by J. M. Cohen, Penguin Classics, 1972.

STARKEY, THOMAS, *A Dialogue Between Reginald Pole and Thomas Lupset*, ed. by K. M. Burton, London, 1948.

STIBLIN, KASPAR, *De Eudaemonensium Republica*, with an introduction and bibliography of the author edited by L. Firpo, Turin, 1959.

VALDÉS, ALFONSO DE, *Diálogo de Mercurio y Carón*, edited and annotated by J. F. Montesinos, Madrid, 1929.

ZUCCOLO, LUDOVICO, *Il Belluzzi ovvero La Città Felice*, ed. Amy A. Bernardy, Bologna, 1929.

— — *La Repubblica d'Evandria e altri dialoghi politici*, ed. R. de Mattei, Rome, 1944.

II. SECONDARY SOURCES

ADAMS, R. P., 'Designs by More and Erasmus for a New Social Order', *Studies in Philology*, Vol. 42, 1945, pp. 132-45.

— — 'The Social Responsibilities of Science in Utopia, New Atlantis and After', *Journal of the History of Ideas*, Vol. 10, 1949, pp. 375-95.

ALLEN, D. C., 'The Degeneration of Man and Renaissance Pessimism'. *Studies in Philology*, Vol. 35, 1938, pp. 202-27.

ALLEN, J. W., *A History of Political Thought in the Sixteenth Century*, London, 1964.

ALLEN, P., 'Medical Education in 17th-Century England', *Journal of the History of Medicine*, 1946, 1: 115-43.

AMES, R., *Citizen Thomas More and His Utopia*, Princeton University Press, 1949.

ARCARI, P. M., *Il pensiero politico di F. Patrizi da Cherso*, Rome, 1935.

ARMYTAGE, W. H. G., *Heavens Below: Utopian Experiments in England 1560-1960*, London, 1961.

— — *Yesterday's Tomorrows: A Historical Survey of Future Societies*, London, 1968.

ATKINSON, G., *The Extraordinary Voyage in French Literature before 1700*, New York, 1920.

— — *Les Nouveaux Horizons de la Renaissance française*, Paris, 1935.

AVINERI, S., 'WAR AND SLAVERY IN MORE'S UTOPIA', *International Review of Social History*, 7 (1962), pp. 260-90.

BARON, H., *The Crisis of the Early Italian Renaissance*, Princeton University Press, 1966.

BATKIN, L. M., 'The Paradox of Campanella', *Diogenes*, 1973, pp. 77-97.

BAUER, H., *Kunst und Utopie: Studien über das Kunst-und-Staatsdenken in der Renaissance*, Berlin, 1965.

BAUMANN, F. L., 'Sir Thomas More', *Journal of Modern History*, Vol. 5, 1932, pp. 605-17.

BEECHER, J. AND BIENVENU R. (EDS.), *The Utopian Vision of Charles Fourier*, Boston, 1971.

BEGLEY, W., *Nova Solyma: An Anonymous Romance Written in the Time of Charles I, Attributed to the Illustrious John Milton*, London, 1902.

BELL, S. G., 'Johan Eberlin von Günzburg's Wolfaria: The First Protestant Utopia', *Church History*, Vol. 36, 1967, pp. 3-20.

BENEDEK, T. G., 'The Image of Medicine in 1500: Theological Reactions to the Ship of Fools', *Bulletin of the History of Medicine*, 1964, pp. 329-42.

BERNERI, M. L., *Journey Through Utopia*, New York, 1971.

BERTANA, E., *Un socialista del cinquecento*, Genoa, 1892.

BIERMAN, J., 'Science and Society in the New Atlantis and Other Renaissance Utopias', *Publications of the Modern Language Association*, Vol. 73, 1963, pp. 492-500.

BLANCHET, L., *Campanella*, New York, 1964.

BLODGETT, E. D., 'Bacon's New Atlantis and Campanella's Civitas Solis: A Study in Relationships', *Publications of the Modern Language Association*, Vol. 46, 1931, pp. 763-80.

BLOOMFIELD, O., *Imaginary Worlds or the Evolution of Utopias*, London, 1932.

BONANSEA, B. M., *Tommaso Campanella: Renaissance Pioneer of Modern Thought*, Washington, DC, 1969.

BOUWSMA, W. J., *Concordia Mundi: The Career and Thought of G. Postel*, Cambridge, Mass., 1957.

— — 'Lawyers and Early Modern Culture', *American Historical Review*, Vol. 78, 1973, pp. 307-18.

BROCKINGTON, C. F., *A Short History of Public Health*, 1956.

BRUBACHER, J. S., *A History of the Problems of Education*, New York, 1966.

BUBER, M., *Paths in Utopia*, London, 1949.

BURKE, P., *Culture and Society in Renaissance Italy 1420-1540*, London, 1972.

BURSTEIN, S. R., 'Care of the Aged in England: From Medieval Times to the End of the Sixteenth Century', *Bulletin of the History of Medicine*, Vol. 22, 1948, pp. 738-46.

BURY, J. B., *The Idea of Progress: an Inquiry into its Origin and Growth*, London, 1928.

BUSH, D., *English Literature in the Earlier 17th Century 1600-1660*, Oxford, 1945.

CAMPBELL, W. E., *Erasmus, Tyndale and More*, London, 1949.

— — *More's Utopia and His Social Teaching*, London, 1930.

CANTIMORI, D., *Eretici italiani del cinquecento: Richerche storiche*, Florence, 1967.

CASPARI, F., *Humanism and the Social Order in Tudor England*, University of Chicago Press, 1954.

CAWLEY, R. R., *Unpathed Waters: Studies in the Influence of the Voyages on Elizabethan Literature*, Princeton University Press, 1940.

CHAMBERS, R. W., *Thomas More*, London, 1945.

CHARLTON, K., *Education in Renaissance England*, London, 1965.

CIORANESCU, A., 'Utopia: Land of Cocaigne and Golden Age', *Diogenes*, No. 75, 1971, pp. 86-117.

CIPOLLA, C. M., *Cristofano and the Plague: A Study in the History of Public Health in the Age of Galileo*, London, 1973.

— — *Public Health and the Medical Profession in the Renaissance*, Cambridge, 1976.

CLARKSON, L. A., *The Pre-Industrial Economy in England 1500-1750*, London, 1971.

COHN, N., *The Pursuit of the Millennium*, London, 1970.

COPEMAN, W. S. C., *Doctors and Disease in Tudor Times*, London, 1960.

CRO, S., 'The New World in Spanish Utopianism', *Alternative Futures*, 1979, Vol. 2, (3): 39-53.

CROCE, B., *Uomini e cose della vecchia Italia*, Bari, 1927.

— — and CARAMELLA, S. (eds.), *Politici e moralisti del seicento*, Bari, 1930.

CROWTHER, J. G., *Francis Bacon*, London, 1960.

CURCIO, C., *Dal Rinascimento alla controriforma: Contributo alla storia del pensiero politico italiano da Guicciardini a Botero*, Rome, 1934.

DAVIS, N. Z., *Society and Culture in Early Modern France* (Eight Essays), London, 1975.

DEBUS, A. G. (ed.), *Science, Medicine and Society in the Renaissance. Essays to Honor Walter Pagel*, New York, 1972.

D'ENTREVES, A. P., *Natural Law*, London, 1970.

DERMENGHEM, E., *Thomas Morus et les utopies de la Renaissance*, Paris, 1927.

DESROCHE, H., 'Messianismes et utopies: note sur les origines du socialisme occidental', *Archives de sociologie de religions*, 1959, 4(8): 31-46.

DONNER, H. W., *Introduction to Utopia*, Uppsala, 1945.

DUHAMEL, P. A., 'Medievalism of More's Utopia', *Studies in Philology*, Vol. 52, 1955.

DUPONT, V., *L'Utopie et le roman utopique dans la littérature anglaise*, Cohors, 1941.

DUVEAU, G., 'Introduction à une sociologie de l'Utopie', *Cahiers Internationaux de Sociologie*, Vol. 9, 1950, pp. 12-24.

EINSTEIN, L., *Tudor Ideals*, London, 1921.

ELIADE, M., 'Paradis et Utopie: Géographie Mythique et Eschatologie', *Eranos Jahrbuch* (32), 1963.

ELLIOTT, J. H., *The Old World and the New 1492-1650*, Cambridge University Press, 1970.

ELLIOTT, R. C., *The Shape of Utopia: Studies in a Literary Genre*, Chicago and London, 1970.

ELTON, G. R., 'An Early Tudor Poor Law', *Economic History Review*, 2nd Ser., Vol. 6, 1953.

— — 'State Planning in Early Tudor England', *Economic History Review*, 2nd Ser., Vol. 13, 1961.

EURICH, N., *Science in Utopia: A Mighty Design*, Harvard University Press, 1967.

FEBVRE, L., *Pour une histoire à part entière*, Paris, 1962.

— — *Le Problème de l'incroyance au XVI siècle: la religion de Rabelais*, Paris, 1942.

FERGUSON, A. B., *The Articulate Citizen and the English Reformation*, Durham, N.C., 1965.

— — 'Renaissance Realism in the "Commonwealth" Literature of Early Tudor England', *Journal of the History of Ideas*, Vol. 16, 1955, pp. 287-305.

FERGUSON, J., *Utopias of the Classical World*, London, 1975.

FIRPO, L., *Bibliographia degle scritti di Tommaso Campanella*, Turin, 1940.

— — 'La città ideale del Filarete', *Studi in Memorio di G. Solari*, Turin, 1954.

— — *Il pensiero politico del Rinascimento e della Controriforma*, Milan, 1966.

— — 'Renaissance Utopianism', in E. Cochrane (ed.), *The Late Renaissance in Italy*, London, 1970, pp. 149-67.

— — *Ricerche Campanelliane*, Florence, 1947.

— — *Gli Scritti di Francesco Pucci*, Turin, 1957.

— — *La stato ideale della Controriforma: L. Agostini*, Bari, 1957.

FISHER, F. J. (ed.), *Essays in the Economic and Social History of Tudor and Stuart England*, Cambridge, 1961.

FRANCK, A., *Reformateurs et publicistes de l'Europe au XVII siècle*, Paris, 1881.

FRANKLIN, A., *La Vie privée d'autrefois — arts et métiers, modes, mœurs, usages des parisiens du XII au XVII siècle*, Paris, 1890.

FUZ, J. K., *Welfare Economics in English Utopias*, The Hague, 1952.

GALDSTON, I., 'Humanism and Public Health', *Bulletin of the History of Medicine*, 1940, pp. 1032-9.

GARIN, E., *L'educazione in Europa 1400-1600*, Bari, 1957.

— — *Italian Humanism, Philosophy and Civic Life in the Renaissance*, Oxford, 1965.

GIBSON, R. W. (ed.), *St. Thomas More: A Preliminary Bibliography of his Works and of Moreana to the year 1750. With a Bibliography of Utopiana Compiled by R. W. Gibson and J. Max Patrick*, Yale University Press, 1961.

GILBERT, F., "The Venetian Constitution in Florentine Political Thought', in N. Rubinstein (ed.), *Florentine Studies*, London, 1968.

GILMORE, M. P., *Humanists and Jurists: Six Studies in the Renaissance*, Cambridge, Mass., 1963.

GRAHAM, H. J., 'The Englishing of English Law', *Moreana*, No. 11, 1966, pp. 25-37.

GRAHAM, R. B. C., *A Vanished Arcadia (The Jesuits in Paraguay 1607-1767)*, London, 1924.

GRAUS, F., 'Social Utopias in the Middle Ages', *Past and Present*, No. 38, 1967, pp. 5-20.

GRENDLER, P. F., *Critics of the Italian World, 1530-1560: Anton Francesco Doni, Nicolò Franco and Ortensio Lando*, University of Wisconsin Press, 1969.

— — 'Utopia in Renaissance Italy: Doni's New World', *Journal of the History of Ideas*, Vol. 26, 1965, pp. 479-94.

GREY, E., *Guevara, a Forgotten Renaissance Author*, The Hague, 1973.

GRUMAN, G. J., 'The Rise and Fall of Prolongevity Hygiene', *Bulletin of the History of Medicine*, 1961, pp. 221-9.

HALE, J., 'A World Elsewhere', in D. Hay (ed.), *The Age of the Renaissance*, London, 1967.

HALL, A. R., 'Science, Technology and Utopia in the 17th Century', in P. Mathias (ed.), *Science and Society 1600-1900*, Cambridge, 1972.

HANSOT, E., *Perfection and Progress: Two Modes of Utopian Thought*, London, 1974.

HAY, D., 'Schools and Universities', *The New Cambridge Modern History*, Vol. II.

HAYDN, E., *The Counter-Renaissance*, New York, 1950.

HERTZLER, J. O., *The History of Utopian Thought*, New York, 1923.

HEXTER, J. H., *More's Utopia: The Biography of an Idea*, Princeton University Press, 1952.

— — 'Thomas More: On the Margins of Modernity', *Journal of British Studies*, I, 1961.

— — 'Utopia and Its Historical Milieu', in *The Complete Works of St. Thomas More*, Vol. 4, Yale University Press, 1965.

— — *The Vision of Politics on the Eve of the Reformation (More, Machiavelli, Seyssel)*, London, 1973.

HIRST, L. F., *The Conquest of Plague*, Oxford, 1953.

HODGEN, M. T., *Early Anthropology in the 16th and 17th Centuries*, Chs. 5, 9, 10, Philadelphia, 1964.

HOWE, G. M., *Man, Environment and Disease in Britain*, New York, 1972.

HUIZINGA, J., *Erasmus and the Age of Reformation*, New York, 1957.

IVES, E. W., 'The Common Lawyers in Pre-Reformation England', *Transactions of the Royal Historical Society*, 5th Ser., Vol. 18, 1968, pp. 145-74.

JOHNSON, R. S., *More's Utopia: Ideal and Illusion*, Yale University Press, 1969.

JONES, H. M., *O Strange New World: American Culture: The Formative Years*, New York, 1964.

JONES, W. R. D., *The Tudor Commonwealth, 1529-1559*, London, 1970.

JORDAN, W. K., *Philanthropy in England, 1480-1660*, London, 1959.

JÜTTE, R., 'Poor Relief and Social Discipline in Sixteenth-Century Europe', *European Studies Review*, Vol. 11, no. 1., Jan. 1981, pp. 25-52.

KATEB, G., *Utopia and Its Enemies*, London, 1963.

KAUTSKY, K., *Communism in Central Europe in the Time of the Reformation*, London, 1897.

— — *Thomas More and His Utopia*, translated by H. J. Stenning, New York, 1959.

— — *et al., Vorläufer des neuren Sozialismus*, Stuttgart, 1895.

KINGDOM, R. M., 'Social Welfare in Calvin's Geneva', *American Historical Review*, Vol. 76, 1971, pp. 50-66.

KRISTELLER, P. O., *Eight Philosophers of the Italian Renaissance*, London, 1965.

— — *Renaissance Thought: The Classic, Scholastic and Humanist Strains*, New York, 1961.

— — and WEINER, P. P. (eds.), *Renaissance Essays*, New York, 1968.

LACHÈVRE, F., *Les Successeurs de Cyrano de Bergerac*, Paris, 1933, pp. 261-9.

LASKY, M. J., *Utopia and Revolution*, Chicago and London, 1976.

LAVEDAN, P., *Histoire de l'urbanisme*, Vol. 2: *Renaissance et temps moderne*, Paris, 1959.

LEANEY, J., 'Medicine in the Time of Shakespeare', *History Today*, Vol. 13, 1963.

LEVIN, H., *The Myth of the Golden Age in the Renaissance*, New York, 1969.

LILJEGREN, S. B., *Studies on the Origin and Early Tradition of English Utopian Fiction*, Uppsala, 1961.

MACNALTY, SIR ARTHUR, *The History of State Medicine in England*, London, 1948.

— — 'Sir Thomas More as Student of Medicine and Public Health Reform', in E. A. Underwood (ed.), *Science, Medicine and History: Essays in Honor of Charles Singer*, London, 1953, pp. 418-36.

MALON, B., *Histoire du socialisme*, Milan, 1879.

MANICARDI, L., 'La repubblica immaginaria di L. Agostini', *La Rassegna*, Vol. 34, 1926, pp. 1-10.

MANNHEIM, K., *Ideology and Utopia*, New York, 1936.

— — 'Utopia', in the *Encyclopaedia of the Social Sciences*, Vol. 15.

MANUEL, F. E. (ed.), *Utopias and Utopian Thought*, London, 1973.

— — and MANUEL, F. P., *French Utopias: An Anthology of Ideal Societies*, New York, 1966.

— — — — 'Sketch for a Natural History of Paradise', *Daedalus*, Winter 1972.

— — — — *Utopian Thought in the Western World*, Harvard University Press, 1979.

MARC'HADOUR, G., *L'Univers de Thomas More: Chronologie critique de More, Erasme et leur époque (1477-1536)*, Paris, 1963.

MARSILI-LIBELLI, C. R., *Anton Francesco Doni: Scrittore e stampatore*, Florence, 1960.

MARTINES, L., *Lawyers and Statecraft in Renaissance Florence*, Princeton University Press, 1968.

MATTEI, R. DE, 'Contenuto ed origini dell'utopia cittadina nel seicento', *Rivista Internazionale di Filosofia del Diritto*, Vols. 9 (1929), 10 (1930), 18 (1938).

— — *La Politica di Campanella*, Rome, 1927.

— — 'Lo Zuccolo, suoi anticipatori ed epigoni', *Rivista Internazionale di Filosofia del Diritto*, Vol. 28.

MITCHELL, B., *Law, Morality and Religion in a Secular Society*, London, 1967.

MONTEGOMERY, J. W., *Cross and Crucible: Johann Valentin Andreae (1586-1654) Phoenix of the Theologians*, The Hague, 1973.

MORGAN, A. E., *Nowhere Was Somewhere: How History Makes Utopias and How Utopias Make History*, Chapel Hill, 1946.

MORELY, H. (ed.), *Ideal Commonwealths*, London, 1886.

MORTON, A. L., *The English Utopia*, London, 1952.

— — 'Utopias Yesterday and Today', *Science and Society*, Vol. 17, 1953.

MUCCHIELLI, R., *Le Mythe de la cité idéale*, Paris, 1960.

MUELLER, W. R., *The Anatomy of Robert Burton's England*, University of California Press, 1952.

MUMFORD, L., *The Story of Utopias*, New York, 1962.

NEGLEY, G., *Utopian Literature: A Bibliography with a Supplementary Listing of Works Influential in Utopian Thought*, Lawrence, Regents Press of Kansas, 1978.

— — and PATRICK, J. M., *The Quest for Utopia: An Anthology of Imaginary Societies*, New York, 1952.

NELSON, W. (ed.), *Twentieth Century Interpretations of Utopia: A Collection of Critical Essays*, Englewood Cliffs, NJ, 1968.

OGLIVIE, C., *The King's Government and the Common Law 1472-1641*, Oxford, 1958.

PÁSZTOR, E., 'La repubblica cristiana di Ottavio Pallavicino', *Rivista di Studi Politici Internazionali*, 1951.

PATRICK, J. MAX, 'Robert Burton's Utopianism', *Philological Quarterly*, Vol. 27, 1948.

PAUL, L., *Sir Thomas More*, London, 1953.

PENROSE, B., *Travel and Discovery in the Renaissance 1420-1620*, New York, 1962.

PHELAN, J. L., *The Millennial Kingdom of the Franciscans in the New World: A Study of the Writings of Gerónimo de Mendieta (1525-1604)*, University of California Press, 1956.

POPPER, K., *The Open Society and Its Enemies*, Vol. I, New York, 1952.

PULLAN, B., *Rich and Poor in Renaissance Venice: The Social Institutions of a Catholic State to 1620*, Oxford, 1971.

RAWSON, E., *The Spartan Tradition in European Thought*, Oxford, 1969.

RITTER, G., *The Corrupting Influence of Power*, Hadleigh, Essex, 1952.

ROBERTS, R. S., 'The Personnel and Practice of Medicine in Tudor and Stuart England', *Medical History*, 1962, 6: 363-82; 1964, 8: 217-34.

ROSEN, G., *A History of Public Health*, New York, 1958.

ROSENAU, H., *The Ideal City: In Its Architectural Evolution*, London, 1959.

ROSS, H., *Utopias Old and New*, London, 1938.

ROSSI, P., *Francis Bacon: From Magic to Science*, London, 1968.

RUSSELL, F. T., *Touring Utopia: The Realm of Constructive Humanism*, New York, 1932.

RUYER, R., *L'Utopie et les utopies*, Paris, 1950.

SADLER, J. E., *J. A. Comenius and the Concept of Universal Education*, London, 1966.

SANFORD, C. L., *The Quest for Paradise: Europe and the American Moral Imagination*, University of Illinois Press, 1961.

SARGEANT, L. T., *British and American Utopian Literature 1516-1975*, Boston, 1979.

SARGENT, L. T., 'A Note on the Other Side of Human Nature in the Utopian Novel', *Political Theory*, Vol. 3, 1975, pp. 88-97.

SCHLANGER, J. E., 'Power and Weakness of the Utopian Imagination', *Diogenes*, 1973, pp. 4-25.

SCHOECK, R. J., 'More, Plutarch, and King Agis: Spartan History and the Meaning of Utopia', *Philological Quarterly*, Vol. 35, 1956, pp. 365-75.

— — 'Rhetoric and Law in Sixteenth-Century England', *Studies in Philology*, Vol. 50, 1953, pp. 110-75.

SCHWOERER, L. G., *'No Standing Armies!' The Anti-army Ideology in 17th Century England*, Johns Hopkins Univeristy Press, 1975.

SÉGUY, J., 'Une sociologie des sociétés imaginées: Monachisme et Utopie', *Annales: Economies, Societies, Civilisations*, Vol. 26, 1971.

SERVIER, J., *Histoire de L'utopie*, Paris, 1967.

SESSIONS, K. C., 'Christian Humanism and Freedom of a Christian: Johann Eberlin von Günzburg to the Peasants', in L. P. Buck and J. W. Zophy (eds.), *The Social History of the Reformation*, Columbus, Ohio, 1972, pp. 137-55.

SIGERIST, H. E., *Landmarks in the History of Hygiene*, London, 1956.

SILVETTE, H., 'Medicine in Utopia', *Bulletin of the History of Medicine*, 1939, pp. 1013-36.

SINGER, C. and UNDERWOOD, E. A., *A Short History of Medicine*, Oxford, 1962.

SKINNER, Q., *The Foundations of Modern Political Thought*, Cambridge, 1978.

SPENS, W. DE, 'les Royaumes d'Utopie', *La Table ronde*, 1962, pp. 168-70.

SPITZ, L. W., 'Johannes Eberlin', in *New Catholic Encyclopedia*, 5: 28-9.

SURTZ, E. L., 'Thomas More and Communism', *Publications of the Modern Language Association*, Vol. 64, 1949.

— — *The Praise of Pleasure: Philosphy, Education and Communism in More's Utopia*, Harvard University Press, 1957.

— — *The Praise of Wisdom*, Chicago, 1957.

TALMON, J. L., *Utopianism and Politics*, London, 1957.

TALMON, Y., 'Pursuit of the Millennium: the Relation between Religion and Social Change', *European Journal of Sociology*, 1962, 3: 125-48.

THIRSK, J. and COOPER, J. P. (eds.), *Seventeenth-Century Economic Documents*, Oxford, 1972.

THOMPSON, D. W., 'Japan and the New Atlantis', *Studies in Philology*, Vol. 30, 1939, pp. 59-68.

THRUPP, S. L. (ed.), *Millennial Dreams in Action: Essays in Comparative Study*, The Hague, 1962.

TOON, P. (ed.), *Puritans, The Millennium and the Future of Israel: Puritan Eschatology 1600-1660*, London, 1970.

TREVOR ROPER, H. R., 'Three Foreigners' in *Religion, the Reformation and Social Change and Other Essays*, London, 1972.

TURNBULL, G. H., *Dury, Hartlib and Comenius*, Liverpool, 1947.

TUVESON, E. L., *Millennium and Utopia: A Study in the Background of the Idea of Progress*, University of California Press, 1949.

Les Utopies á la Renaissance, Colloque International (1961), Travaux de l'institut pour l'étude de la Renaissance et de l'humanisme, Brussels and Paris, 1963.

WALKER, W. B., 'Luigi Cornaro, A Renaissance Writer on Personal Hygiene', *Bulletin of the History of Medicine*, 1954, pp. 525-34.

WALSH, C., *From Utopia to Nightmare*, London, 1962.

WARREN, F. B., *Vasco de Quiroga and his Pueblo Hospitals of Santa Fe*, Washington, DC, 1963.

WATSON, F. (ed.), *Vives and the Renascence Education of Women*, London, 1912.

WATSON, F., *Vives: On Education*, Cambridge, 1913.

WEBSTER, C., *The Great Instauration: Science, Medicine and Reform, 1626-1660*, London, 1975.

— — 'The Authorship and Significance of Macaria', *Past and Present*, No. 56, 1972.

WEISINGER, H., 'Ideas of History During the Renaissance', *Journal of the History of Ideas*, Vol. 6, 1945, pp. 400-35.

WHITE, H. C., *Social Criticism in Popular Religious Literature of the Sixteenth Century*, New York, 1944.

WILLEY, B., *The Seventeenth Century Background*, London, 1949.

WILLIAMS, A., 'A Note on Pessimism in the Renaissance'. *Studies in Philology*, Vol. 36, 1939, pp. 243-6.

WILSON, F. P., *The Plague in Shakespeare's London*, OUP, 1963.

WOODBINE, G. E., 'The Language of English Law', *Speculum*, Vol. 18, 1943, pp. 396-434.

WOODWARD, W. H., *Vittorino da Feltre and Other Humanist Educators*, Cambridge, 1897.

— — *Studies in Education during the Age of the Renaissance*, Cambridge, 1906.

YATES, F. A., *The Rosicrucian Enlightenment*, London, 1972.

— — *Giordano Bruno and the Hermetic Tradition*, London, 1971.

— — *The Art of Memory*, London, 1966.

ZAVALA, S., *Sir Thomas More in New Spain: A Utopian Adventure of the Renaissance*, London, 1955.

— — *New Viewpoints on the Spanish Colonization of America*, ch. 10: 'Spanish Colonization and Social Experiments', University of Philadelphia Press, 1943.

ZILSEL, E., 'The Genesis of the Concept of Scientific Progress', *Journal of the History of Ideas*, Vol. 6, 1945, pp. 325-50.

INDEX